NUMERICAL SIMULATIONS: WELDING INDUCED DAMAGE STAINLESS STEEL 15-5PH

WU TONG

CHINA ARCHITECTURE & BUILDING PRESS

图书在版编目（CIP）数据

基于相变场的金属材料损伤研究及工程应用 =
NUMERICAL SIMULATIONS: WELDING INDUCED DAMAGE
STAINLESS STEEL 15-5PH : 英文 / 吴通著. -- 北京 :
中国建筑工业出版社, 2025. 4. -- ISBN 978-7-112
-30993-1

Ⅰ. TG14

中国国家版本馆 CIP 数据核字第 2025Y2X375 号

责任编辑：毕凤鸣
文字编辑：孙晨淏
责任校对：王　烨

NUMERICAL SIMULATIONS: WELDING INDUCED DAMAGE
STAINLESS STEEL 15-5PH
WU TONG

*

中国建筑工业出版社出版、发行（北京海淀三里河路 9 号）
各地新华书店、建筑书店经销
国排高科（北京）人工智能科技有限公司制版
建工社（河北）印刷有限公司印刷

*

开本：787 毫米 × 1092 毫米　1/16　印张：12½　字数：312 千字
2025 年 5 月第一版　　2025 年 5 月第一次印刷
定价：**76.00** 元

ISBN 978-7-112-30993-1
（44663）

如有内容及印装质量问题，请与本社读者服务中心联系
电话：（010）58337283　　QQ：2885381756
（地址：北京海淀三里河路 9 号中国建筑工业出版社 604 室　邮政编码：100037）

Abstract

The objective of this study is the prediction of damage and residual stresses induced by hot processing which leads to phase transformation in martensitic stainless steel. This study firstly concerns the modelling of the damage of material induced by a complex history of thermoelastoplastic multiphase in heat-affected-zone (HAZ) of welding. In this work, a two-scale mode of elastoplastic damage multiphase was developed in the framework of thermodynamatics of irreversible process. The constitutive equations are coupling with ductile damage, elastoplasticity, phase transformation, and transformation plasticity. Besides, a damage equation was proposed based on Lemaitre's damage model in the framework of continuum damage mechanics.

The experiments of 15-5PH were implemented for the identification of phase transformation, transformation plasticity and damage models. Tensile tests of round specimens were used to identify the parameters of the damage model as well as mechanical behaviours at various temperatures. Tests of flat notched specimens were designed to provide the validation of the damage model and strain localization using three-dimensional image correlation technologies. In addition, microscopic analysis was performed to provide microstructure characterization of 15-5PH and to discover the damage mechanism.

Finally the numerical simulation was performed in the code CAST3M® of CEA. On the one hand, numerical verification of the flat notched plates was implemented and compared with experimental results. On the other hand, we used the two-scale model including phase transformation, transformation plasticity and damage to simulate the level of residual stresses of a disk made of 15-5PH metal heated by laser. The internal variables, such as strain, stress, damage, were successfully traced in the simulation of two-scale model. The simulation results showed the transformation plasticity changes the level of residual stresses and should not be negligible; damage decreases about 8 percent of the peak value of residual stresses on upper surface of disk.

KEYWORDS: Damage, Welding, Phase Transformation, Numerical Simulation, Transformation Plasticity, Experiment

Contents

Nomenclature

$\Delta\varepsilon_{\gamma-\alpha}^{T_{ref}}$	difference of compactness of phase α compared to phase γ at T_{ref}
z_i	volume proportion of phase i
ε_i	total microscopic strain of phase i
ε_i^e	elastic microscopic strain of phase i
ε_i^p	plastic microscopic strain of phase i
ε_i^{thm}	thermo metallurgical microscopic strain of phase i
E	Young's modulus
E^c	classical macroscopic strain
E^e	elastic macroscopic strain
E^p	plastic macroscopic strain
E^{tot}	total macroscopic strain
E^{pt}	phase transformation plastic strain
R_i	isotropic strain hardening
$\underline{X_i}$	kinematic strain hardening
F_i	yield function of phase i
p_i	equivalent plastic strain
$\underline{\alpha_i}$	internal variable associated with kinematic hardening
$\dot{r}_i$	internal variable associated with isotropic hardening
φ^D	damage dissipation potential
T	temperature
T_{ref}	reference temperature
D_i	damage variable of phase i
Σ	macroscopic stress
$\underline{\sigma_i}$	microscopic stress of phase i
σ_γ^y	yield strength of phase i
σ_{eq}	equivalent stress
S	deviator of stress of phase α
H	Hooke's operator
$\alpha_i(T)$	dilatation coefficient of phase i (depend on temperature T)

$\bar{\alpha}_i$	dilatation coefficient of phase i (mean value)
M_s	martensite start temperature
A_{c_1}	austenitic transformation start temperature
A_{c_3}	austenitic transformation finish temperature
β	coefficient depend on material in phase transformation model
ε_{eq}^{pt}	equivalent plastic strain induced by phase transformation
ε^{pt}	transformation induced plasticity
p_D	plastic strain to threshold damage
p_R	plastic strain at failure
D_0	threshold damage
D_c	damage at failure
κ	damage exponent depend on material
I	unit tensor

Chapter 1 Introduction

The research on welding, which includes thermal, metallurgical and mechanical phenomena, is a both old and new topic because many studies on the subject were started or developed many years ago and some are still going on.

The welding processes generally leaves residual stresses in the weld and its vicinity vicinity, since the solidification and contraction of molten material during cooling are constrained by the surrounding metal. The existence of such residual stresses can significantly affect subsequent lifetime by augmenting or impeding fatigue or failure events. The performance in service of parts that have been subjected to welding or heat treatment depends on the residual stress state of the structure. Consequently, for an accurate assessment of engineering lifetimes, there is a need to determine residual stress profiles. From an experimental point of view, a variety of techniques such as neutron, X-ray and synchrotron X-ray diffraction, hole drilling, curvature measurements and instrumented indentation are available to measure residual stresses[143][97][177].

Although experimental techniques provide practical solutions for measuring residual stresses, numerical methods are necessary and should be emphasized in industrial engineering. This is because design and calculation are no longer separated procedures and the calculation nowadays is already involved with the design process. Nowadays, the industrial needs in diverse fields such as aeronautics, nuclear power and automobiles impose increasingly high constraints in order to reduce manufacturing costs and to increase the reliability of parts. Control of the behavior and the rupture of mechanical structures becomes paramount with the development of simulation technology. Consequently, numerical simulation leads to various benefits for industry:

➢ Simulation should help avoid long test periods.

➢ Virtual techniques lead to maximization of performances and optimization of processes.

➢ Computational approaches provide additional insights into the experimental data and offer design capabilities.

➢ Simulation could predict and then prevent cracks or further damage induced by welding process.

For these reasons numerical simulation plays an increasingly dominant role in certain fields of industry. It is well known that the finite element method (FEM) is a very effective approach to calculating residual stresses and damages, and increasing applications emerge in various research and industrial fields with the development of computational or electronic technology. In particular, in the field of metal hot process, virtual process becomes more popular in many institutes and enterprises to predict the results and tune the parameters before implementing the actual experiments or manufacture processes.

The history of finite element simulations of the thermal and mechanical behaviour during welding can be traced back to the 1970s[191][94][75][166][9]. Computational Welding Mechanics (CWM) is concerned with the thermo-mechanical response as well as the changes in material properties due to the thermal cycles in welding. The three most important conferences in CWM are Numerical Analysis of Weldability[2], Trends in Welding Research[1] and Modelling of Casting, Welding, and Advanced Solidification[3]. Lindgren[130] introduced different modelling aspects of the application of the finite element method to predict the thermal, material and mechanical effects of welding. Besides, he introduced the complexity of welding simulation, material modelling of welding and computational strategies of welding simulation[127][128][129].

In general, the prediction of welding residual stresses can be classified as following: statistical and empirical curves or formula, analytic method, inherent strain method[192][139][203] and thermo viscoelastoplastic finite element analysis. The first two methods are based on many simplifications and hypotheses, which lead to limited applications in industry. The concepts of inherent strain and stress were developed in Japan decades ago. This method meets the requirements of predicting welding residual stresses of carbon steels. The thermo elasto-plastic finite element analysis seems to be the most promising approach, and it has been well developed by many researchers[145][162][184][46]. In our study, we used this method to implement the models.

Therefore, our study aims to develop a numerical welding model under phase transformation including damage prediction. A successful model needs to consider thermal, mechanical and metallurgical phenomena (Figure 1.1) that are involved during the welding process. This coupled problem is usually solved by using finite element (FE) models where the thermal, mechanical and metallurgical analyses are either performed simultaneously or staggered. In prior literature[98][89][7][49], coupling was studied between thermics and mechanism, between thermics and metallurgy as well as the influence of metallurgy on mechanics.

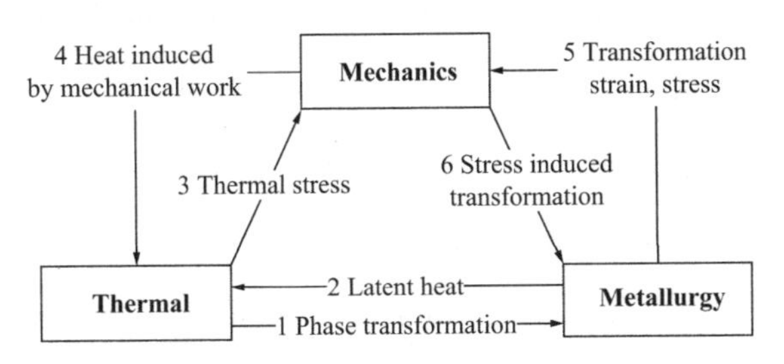

Figure 1.1 Coupling mechanisms[98]

When a model has been created, how could it be evaluated? For a successful model with application to realistic materials, one should not neglect the following two aspects: first, calibrations of parameters; second, model validations. Validation is the process during which the accuracy of the model is evaluated by comparison with experimental results whereas calibration to match the parameters with some benchmark (Figure 1.2). Thus, the same measurements should not be used for both calibration and validation purposes. An adequate model thus should have sufficient accuracy for such purpose. Here, verification is performed by assuming that the finite element model is "correct" with respect to the conceptual model and qualification is a similar process but between conceptual model and reality.

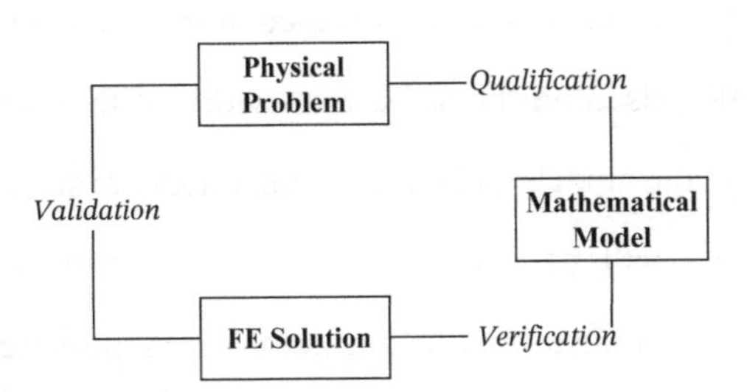

Figure 1.2 Validation and verification in finite element modeling[144]

Our work is a part of Project INZAT, which is dedicated to studying simulation and experimenting of welding of metallic materials. The project is carried out in our laboratory (LaMCoS, INSA de Lyon) and lasted more than 10 years, and is supported by various industrial enterprises of France: AREVA, EDF, EADS and ESI. Some recent work of INZAT are shown in the references[53][62][63][193][197][199][207][208][209].

Our study extends M. Coret's work that is dedicated to experimental study and numerical simulation of the welding residual stress of 16MND5 and SA533[45][46][47][48]. Another previous study could be traced back to the thesis work of M. Martinez on weld joint of 16MND5-INCONEL[135]. In our study, the numerical model incorporates more factors such as damage and is applied for 15-5PH instead of 16 MND 5 stainless steel. The metal 15Cr-5Ni (a martensitic precipitation hardening stainless steel) which is examined in our research undergoes the phase transformation of martensite to austenite during heating or vice verse during cooling of welding process. In the model, we will take the three most important phenomena into account, namely damage, phase transformation and transformation plasticity. In previous studies, much work was devoted to the study of phase transformation and TRansformation Induced Plasticity (TRIP). Several authors have developed constitutive models for the mechanical behavior of dual-phase TRIP steels[112][117][146][59]. Leblond et al.[116][117] consider that in a mixture of two phases with different specific volumes, volume incompatibilities generate microscopic plasticity in the weaker phase, namely the one that has the lower yield stress. He then proposed a proper TRIP model. On the other hand, the mechanism of damage has been extensively studied in the literature from the metallurgical to the modeling point of view, such as Lemaitre-Chaboche damage model for elastoplastic material[124][125], Rousselier ductile fracture model[163] and Gurson's ductile fracture model[87]. The literature concerning this topic is too broad to make complete quotations[18][168][134][104][158][29][86].

However, there is little evidence for the existence of models couple both damage and phase transformation (even TRIP) based on the same framework of thermodynamics. This implies to cope with a diversity of damage models and increase complexity to standard constitutive equations. In our study, a new model will be developed. This model contains three main ingredients, continuum damage mechanics, transformation plasticity and multiphase behavior. This model will be applied to calculate the welding residual stresses of 15-5PH under phase transformation and damage conditions. As mentioned above, calibration and validation are necessary for practical applications. Thus, detailed processes of calibration and validation for the damage and phase transformation models, which are two main difficulties in numerical simulation, will be given after the descriptions of two models in the thesis. Next, the numerical simulations of several cases will be provided to illustrate the verification and validation of the constitutive models.

In summary, the presented study will be performed in the framework of the thermodynamics of damage. Its guideline principles are to combine a detailed experimental analysis of the considered materials and of their specific damage mechanisms, a realistic modeling of these mechanisms and the implementation of these models into a numerical simulation of the response of the structural components under investigation.

Chapters after the introduction section are summarized as following:

Chapter 2 presents the specific material studied and its metallurgical aspect. Phenomena of phase transformation, various models of phase transformation, TRIP models, several mechanical models of multiphase are introduced in details.

Chapter 3 gives an introduction to Continuum Damage Mechanics (CDM). Definition of damage variables and constitutive equations of damage evolution in the framework of thermodynamics are introduced.

The above two chapters are the introductions of some existing theories and models related to our study, which are selected from numerous literatures.

Chapter 4 is dictated to develop the coupling between damage and phase transformation, and then to provide the coupled constitutive equations in mesoscopic scale and two-scale in the framework of thermodynamics. In addition, a damage equation was proposed based on the Lemaitre's damage model.

Chapter 5 focuses on experimental study and parameter identification of the stainless steel 15-5PH. These parameters include classical material properties, parameters of phase transformation, transformation plasticity and damage models. The means of experiments include the circularly tensile tests of round bars, the tensile tests of flat notched specimens and the micro observations. The three dimensional digital image correlation (DIC) is used to study strain localization on surface of notched specimen.

In **Chapter 6**, the numerical simulation is implemented in CEA CAST3M® finite element code[27]. The numerical verification of the flat notched bars is performed to compare with experimental results. Another example is the simulation of disk heated by laser, in which the phase transformation, transformation plasticity and damage are included in two-scale model. The damage and residual stress will be predicted in macro, meso and two-scale models.

At the end, some concluding remarks are found in **Chapter 7**.

Chapter 2 Martensitic Stainless Steel and Solid-state Phase Transformation

2.1 Introduction

Phase transformations often play a dominant role in the modelling of certain thermomechanical problems. General concept of thermodynamic and phase transformation is presented in[148] within the framework of solid-state physics. This chapter is dedicated not only to the mechanical behaviour and constitutive modelling but also to physical and metallurgical aspects. Introductions to concepts and models of transformation-induced volumetric strain and Transformation-Induced Plasticity (TRIP) will be found in this chapter.

2.2 Martensitic stainless steel and 15-5PH

There are three types of martensitic stainless steels: traditional (Cr13 is typical one), low-carbon and super (or soft) martensitic stainless steels. The traditional type contains 14-17wt(Cr)%[1] and 1-2 wt(C)%. Despite of its good toughness and hardness, the traditional type of martensitic stainless steels has limited application and it is thus replaced by new materials due to its low ductility and poor welding ability. The low-carbon martensitic stainless steels are melted through adding small quantity of a alloying element (e.g. Mo, Ti) and reducing the carbon content. This improves the workability and gains good corrosion resistance. Therefore, this type of martensitic stainless steel is widely applied in hydro-turbines and water pumps. The supermartensitic stainless steel with super low-carbon and low nitrogen has been developed during recent years. Its nickel content is restricted between 4% and 7%, and some alloying

1 wt( )% means weight percent.

elements (e.g. Mo, Ti, Si and Cu) are added. This type of martensitic stainless steel is well applied in pipe and oil industries. Some typical low-carbon and super martensitic stainless steels are given in Table 2.1.

Another important type of martensitic steel is a martensitic precipitation hardening stainless steel, such as 17-4PH, 15-5PH etc. Table 2.2 lists chemical integrations of some typical kinds of martensitic precipitation hardening stainless steels. The alloy investigated in this research is 15-5PH stainless steel (AMS5659), a kind of martensitic precipitation hardening stainless steels steel. It offers high strength and its common applications are components of aerospace including valves, shafts, fasteners, fittings and gears. The position of 15-5PH in metal family is shown in Figure 2.1. The specific 15-5PH which we used in experiment is H1025 (Condition H1025) which means that it is precipitation of age hardened and heat solution-treated material at specified temperature (1025°F)+/−15°F for 4 hours and air cooling. The element compositions of 15-5PH H1025 are shown in Table 2.3 (as provided by the factory). In term of microscopic scale, this steel is martensite at low temperature whereas it is in austenitic state at high temperature. The microscopic observation of 15-5PH in ambient temperature is given in Figure 2.2.

Some typical kinds of low-carbon and super martensitic stainless steels[36] Table 2.1

Type	Standard	C	Mn	Si	Cr	Ni	Mo	Others
ZG0Cr13Ni4Mo (China)	JB/T 7349-94	0.06	1.0	1.0	11.5-14.0	3.5-4.5	0.4-1.0	N,S: 0.03
ZG0Cr13Ni5Mo (China)	JB/T 7349-94	0.06	1.0	1.0	11.5-14.0	4.5-5.5	0.4-1.0	N,S: 0.03
CA-6NM (USA)	ASTM A734/ ASTM-1998	0.06	1.0	1.0	11.5-14.0	3.5-4.5	0.4-1.0	N,S: 0.03
Z4 CND 13-4-M (France)	AFNOR NF A32-059-1984	0.06	1.0	0.8	12.0-14.0	3.5-4.5	0.7	N,S: 0.03
Z4 CND 16-4-M (France)	AFNOR NF A32-059-1984	0.06	1.0	0.8	15.5-17.5	4.0-5.5	0.7-1.5	N,S: 0.03
12Cr-4.5Ni-1.5Mo (France)	CLI Company	0.015	2.0	0.4	11.0-13.0	4.0-5.0	1.0-2.0	N: 0.012 S: 0.002
12Cr-6.5Ni-2.5Mo (France)	CLI Company	0.015	2.0	0.4	11.0-13.0	6.0-7.0	2.0-3.0	N: 0.012 S: 0.002

Note: Single value means maximum.

Chemical compositions of some typical martensitic precipitation hardening stainless steels[36] Table 2.2

	C	Mn	Si	Cr	Ni	Cu	Ti	Nb + Ta
17-4PH	0.07	1.00	1.00	15.5-17.5	3.0-4.0	3.0-4.0	—	0.15-0.45

Table 2.2 (Continued)

	C	Mn	Si	Cr	Ni	Cu	Ti	Nb + Ta
15-5PH	0.07	1.00	1.00	14.0-15.5	3.5-5.5	2.5-4.5	—	0.15-0.45
13-8Mo	0.05	0.10	0.10	12.25-13.25	7.5-8.5	—	Al 0.9-1.35	Mo 2.0-2.5
AM362	0.05	0.50	0.30	14.0-15.0	6.0-7.0	—	0.55-0.90	—
AM363	0.05	0.30	0.15	11.0-12.0	4.0-5.0	—	0.30-0.60	—

15Cr-5Ni alloy is readily machined under both solution treated and various age-hardened conditions. In the solution-treated condition, it performs similarly to stainless Types 302 and 304. The machinability will improve as the hardening temperature is increased. 15Cr-5Ni can be satisfactorily welded by the shielded fusion and resistance welding processes. Normally, welding in the solution-treated condition has been satisfactory; however, where high welding stresses are anticipated, it may be advantageous to weld in the overaged (H 1150) condition. Usually, preheating is not required to prevent cracking. If welded in the solution-treated condition, the alloy can be directly aged to the desired strength level after welding. However, the optimum combination of strength, ductility and corrosion resistance is obtained by solution treating the welded part before aging. If welded in the overaged condition, the part must be solution treated and then aged[96].

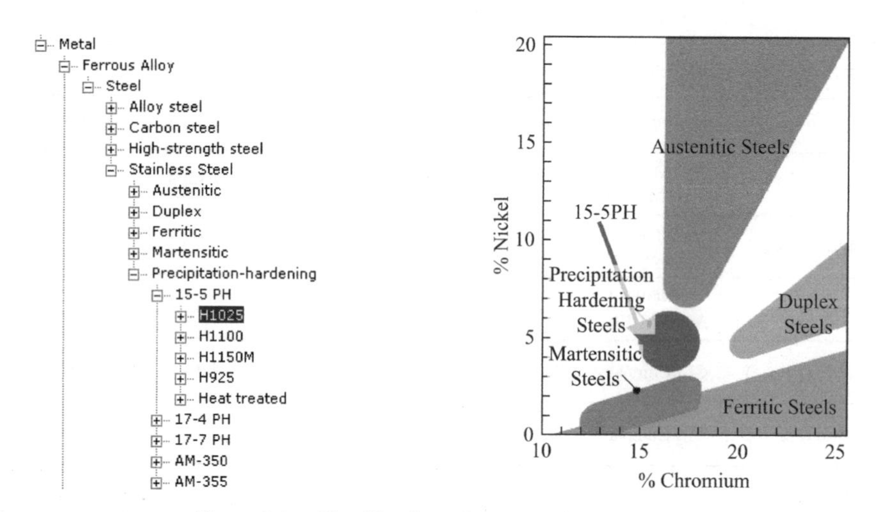

Figure 2.1 Classification of 15-5PH in metal family

Chemical compositions of 15-5PH stainless steel (wt%) Table 2.3

C	SI	Mn	P	S	Cr	Mo	Ni	Cu	Al	N	NB	Fe
0.030	0.40	0.66	0.020	0.001	15.44	0.05	4.50	3.16	0.013	0.0247	0.292	Balance

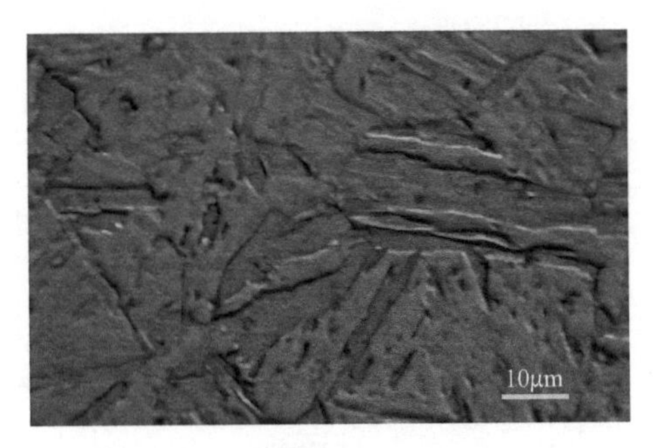

Figure 2.2 Microscopic observation (100 X) of 15-5PH steel

2.3 Phase transformation of martensitic steel

At high temperatures (over typical temperature of 750°C), steel atoms are arranged according to a crystal lattice form known as austenite (FCC structure, Figure 2.3). As steel cools, austenite changes into pearlite, ferrite, bainite or martensite, which have a different lattice structure (BCC structure, Figure 2.3). A schematic diagram of phase transformation of martensitic steel under heating and cooling conditions is given in Figure 2.4. A transition like this, from one type of lattice to another, is referred to as a phase transformation. Solid-state phase transformation causes the macroscopic geometric change because these different types of crystalline structures have different densities, which is the so-called transformation-induced volumetric strain. During a process of welding, the transformations of phase in a solid state, which are the consequences of the imposed thermomechanical loading, result from the combination of the change of crystal lattice of the iron and the displacement of the atoms of aqueous solution. The crystalline states are functions of the temperature, the level and the stress type. The transformations of phase of steels are often studied by using the system of iron-carbon. Figure 2.5 shows the equilibrium diagram for combinations of carbon in a solid solution of iron.

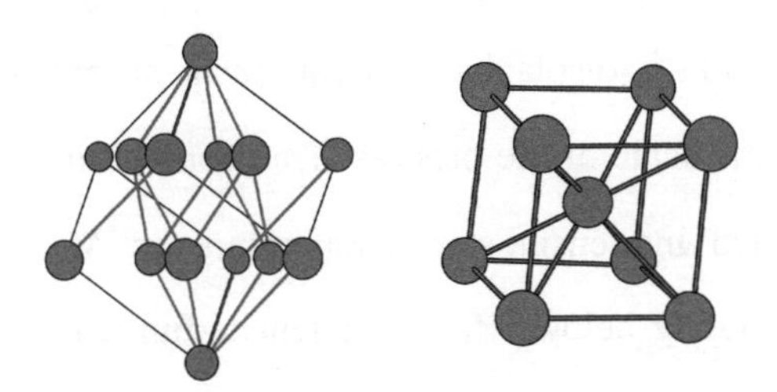

Figure 2.3 Crystal lattice structures of FCC and BCC

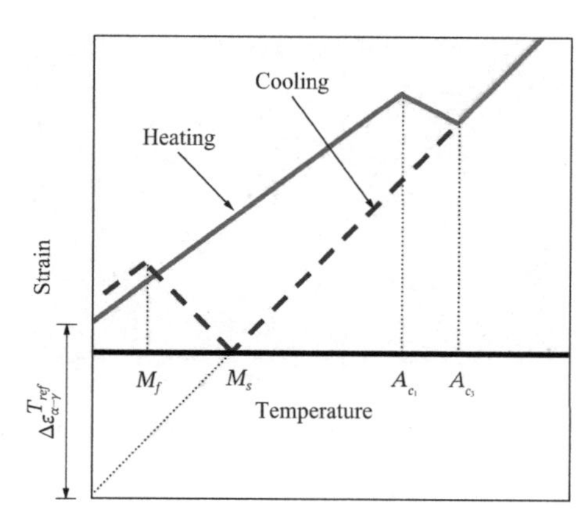

Figure 2.4 Schematic diagram of phase transformation under heating and cooling conditions

In steels, austenite can transform to ferrite, pearlite and martensite either by a reconstructive or by a displacive mechanism.

In the case of reconstructive nucleation, it is formed in the mother phase starting from local inhomogeneousness, an embryo with the composition and the crystalline structure of the new phase. The partition of the aqueous solutions to the interface between the mother phase and the embryo is supposed near to where thermodynamic balance is. In order that the transformation is carried out, it is necessary that the embryo reaches a critical size, and there is a mechanism of growth of the new phase to complete the transformation, in particular of the transport of atoms by diffusion.

In the case of displacive nucleation, it appears in the austenite of local configurations of dislocation which, by slip in a correlated way, causes the displacement of the atoms from their sites on the reticulum of the mother phase to new nearby sites which have the configuration of the crystal lattice of the new phase. The growth of the germ is carried out in the form of slats or of plates, without transport of matter at long distances compared to the interatomic distances. In steels, this type of nucleation intervenes only when, following a forced cooling, the transformation of austenite is carried out at a temperature quite lower than the A_{c_3} temperature.

Martensitic transformation is displacive and can occur at temperatures where diffusion is inconceivable within the time scale of the process. Transformation starts only after cooling to a particular temperature called martensite-start temperature or M_S. The fraction transformed increases with the under cooling below M_s. A martensite-finish temperature or M_f is usually defined as the temperature where 95% of the austenite has decomposed. Unlike M_s, M_f has no fundamental significance.

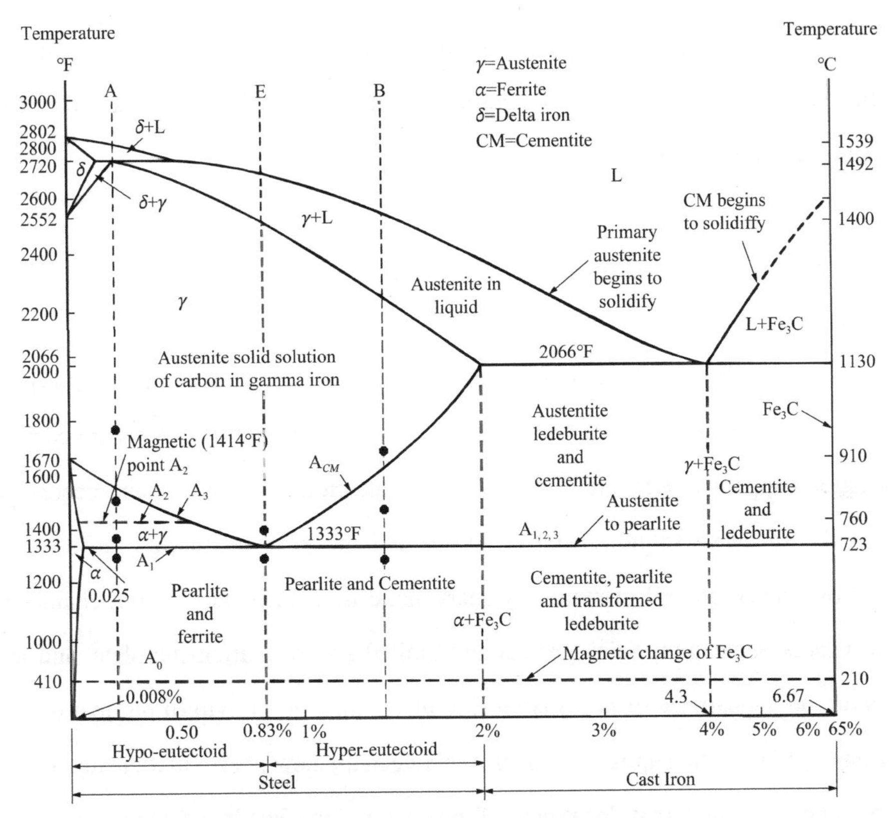

Figure 2.5 Fe-Fe$_3$ C phase diagram[136]

2.3.1 Phase transformation of austenite during heating

When the austenitization is processed at a sufficiently slow rate of heating under conditions of near stably thermodynamic balance, and the initial and end temperatures of transformation are respectively noted A_{c_1} and A_{c_3} (Figure 2.4). When the heating rate is faster, and the initial and end temperatures of transformation are moved towards higher values noted A_{c_1}' and A_{c_3}' and the beach of transformation is extended. Thus, the heating rate of temperature influences the kinetics of transformation and the points of beginning and end of transformation.

During the austenitization, two other parameters play an essential role: the maximum temperature reached and duration period at this temperature. These two parameters determine the parameter of austenitization which leads to a balance between time and the temperature. The higher value of this parameter means the larger size of the austenitic grain. The austenitic size of grain has an influence on the transformations during cooling. All the transformations carry out nucleation and growth. Nucleation is done preferably on discontinuities of the reticulum and for this reason, the grain boundaries represent a privileged place. Modifying the size of the austenitic grain can modify the spatial distribution of the grain boundaries and thus influence the kinetics of transformation during cooling.

2.3.2 Phase transformation of martensite during cooling

The martensitic transformation occurs very far from thermodynamic equilibrium. When austenite is cooled very quickly, carbon does not have time to diffuse and ferrite cannot appear. Transformation happens at temperatures well below eutectoid temperature and occurs extremely fast, i.e. at the speed of sound. Just like ferrite and the pearlite, martensite is formed by a mechanism of nucleation and growth. The martensitic transformation is with characteristics of instantaneousness nucleation, quick growth and limited size of martensitic plate, and its process depends more on appearance of new martensitic plates than on growth of old martensitic plates. New martensitic plates often appear boundaries between the new phase (martensite) and old phase (austenite), and it means that locations of nucleation are inside of grains, not like phase transformation of diffuse type, where nucleation occurs on the grain boundaries (Figure 2.6). During martensitic transformation, there is no diffusion, and the transformation progresses through local atomic rearrangement. It stops transforming the rest austenite when the supplied energy becomes insufficient. In ferritic steels, it is in particular of the martensitic transformation and to a less extent about the bainitic transformation. Under anisothermal conditions, for fast rates of cooling, the ferritic diffusion transformation is stopped (or cannot occur). Contrary to the bainitic transformation, the martensitic transformation occurs in higher cooling rate. However, in some specific martensitic steels, when austenite is cooled, there is only martensitic transformation occurs in low temperature, whereas no phase transformation happens in high temperature or mesothermal state[211][157].

In steels, martensite exists in many forms, according to different morphological characteristics and sub-structure, namely: lath martensite, flake martensite, butterfly shaped martensite, thin sliced martensite and thin plate martensite, etc. The lath martensite and flake martensite are often the cases. Usually, the former appears with a substructure of high-density dislocations, and the substructure of the latter is twin crystal.

Martensite (α') is hindered by austenite grain boundaries whereas allotriomorphic ferrite (α) is not

Figure 2.6 Schematics for the formation of martensite plates

2.3.3 Main factors of influence on transformation

Thermal effect

In terms of thermal effect on phase transformation, there are four factors that should be taken into account: heating rate, highest temperature reached, duration of high temperature stay and cooling rate.

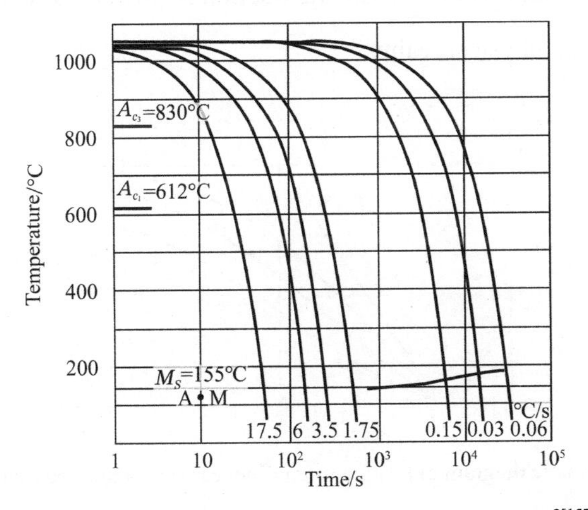

Figure 2.7 CCT diagram of martensitic steel Fv520(B)[2][157]

The phase transformation strongly depends on the cooling rate and the composition of alloy elements. Two types of diagrams are used by the researchers of the heat treatments to represent these transformations simply: Time-temperature transformation (TTT) diagrams are obtained by fast cooling of austenite then maintenance at constant temperature; continuous cooling transformation (CCT) diagrams represents the transformations during cooling at constant speed.

It is known that the decomposition of austenite happens in anisothermal condition and more

2 Fv 520 (B)- w(%): 0.05C, 0.40Si, 0.80Mn, 14.50Cr, 5.50Ni, 1.80Cu, 1.70Mo, 0.35Nb, 0.007P, 0.006S

or less coarse mixture of ferrite-cementite, bainite and martensite according to the speed of cooling. The physical mechanism during the nucleation of the new phases depends on the temperature of start transformation, which is a function of the cooling rate. Higher speed of cooling, the initial temperature of transformation is lower.

In martensitic steel, there are only martensite and austenite during heating and cooling, so these three factors have no influence on the types of phase transformation (Figure 2.7). However, the thermal parameters affect to the phase stability and properties. Generally speaking, the increasing of heating rate can loss the stability of austenite, and make it more difficult to absorb carbonides into austenite. Imcreasing the top temperature increases the grain size of austenite and the stability of austenite.

The heating rate influence on the austenite transformation can be explained in Figure 2.8. For a very slow heating rate, there is enough time to make the austenite fraction to reach equilibrium state for each temperature. For an increasing heating rate, it becomes more and more far away equilibrium state, and the increasing of austenite fraction is slower and with hysteresis effect. This is well in accordance with physical reality.

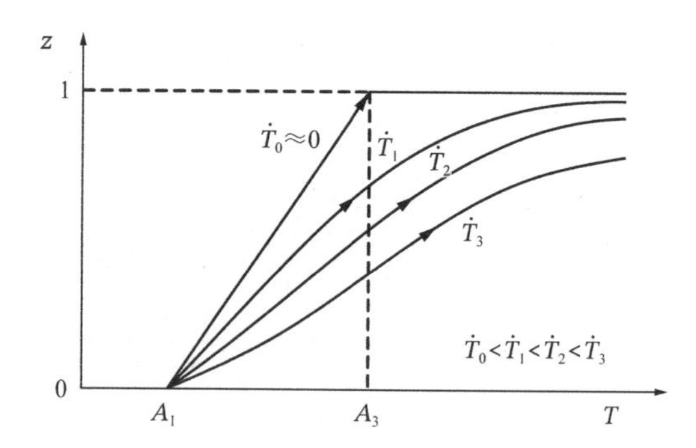

Figure 2.8 Schematic diagram of heating rate influence on austenite phase transformation[114]

Grain size of austenite

The austenitic transformation as well as others' phase-changes processes through nucleation and growth. Usually the nucleation of austenitic transformation occurs preferably on discontinuities of texture and for this reason, the grain boundaries present a preferred site. The size of grain of austenite plays an important role in the transformations during cooling. Several works[44][135], resulting from the literature, evaluated the influence of the size of austenitic grain on the transformations during cooling. These studies show the grain sizes modify not the thermodynamic equilibrium but the kinetics.

In ferrite steels, it can not modify the thermodynamic equilibrium but change both the kinetics of phase transformation and the morphologies of phases formed. However, in supermartensitic stainless steels, the grain size of austenite does not affect the types of phase transformation because martensite is only phase of output of phase transformation (retained austenite possible) during cooling, but it changes the rate of nucleation and growth, and temperature of transformation to some extent.

Percentage of carbon and other elements of addition

Supermartensitic stainless steels and martensitic precipitation hardening stainless steels are essentially alloys based on iron but containing chromium and nickel. They owe their name to their room temperature martensitic microstructure. To understand the metallurgy of this family of steels, attention is paid to the effects that the key alloying elements have on phase stability and properties.

- Carbon and nitrogen

The transformations of phase are related to the possibility of diffusing carbon, and the carbon concentration modifies the transformations of phase. Carbon and nitrogen are strong austenite stabilisers in Fe-Cr alloys. More percentage of carbon leads to more ferrite to be formed in the ferrite steel. However, in supermartensitic stainless steels, the carbon and nitrogen have limited influence, because their contents have to be kept as low as possible, about 0.01 wt%. This is because martensite hardness increases sharply with carbon concentration and therefore raises the probability of sulfide stress corrosion-cracking and hydrogen-induced cold-cracking.

- Chromium

Chromium is a ferrite stabiliser. Low carbon Fe-Cr stainless steels have a ferritic or martensitic, possibly semi-ferritic, microstructure depending on composition. When the chromium content is below approximately 12 wt%, it is possible to obtain a martensitic microstructure since the steel can be made fully austenitic at elevated temperatures. Such steels solidify as δ-ferrite and are completely transformed to austenite (γ) at high temperature, followed by relatively rapid cooling to transformation into non-equilibrium martensite. A chromium content greater than approximately 14 wt% gives a completely ferritic stainless steel over the whole temperature range corresponding to the solid state and hence cannot be hardened on quenching. Between the austenite phase field and the fully ferritic domain, there is a narrow range of compositions which defines the semi-ferritic alloys, with a microstructure consisting partly of δ-ferrite which remains unchanged following solidification, the remainder being martensite (Figure 2.9).

*- **Nickel***

Austenite can be stabilised using substitutional solutes. Nickel has the strongest effect in this respect (Figure 2.10) and also a tendency to improve toughness. Nickel also has influence on Ms (martensite-start temperature) as Figure 2.11. Cr, Ni and Mo concentration has influence on boundaries of the austenite, ferrite and martensite phases (Figure 2.12).

In addition, other elements have their own functions in metallurgy, for example, manganese and copper are austenite stabilising elements whereas silicon and titanium are ferrite stabilising elements. For the sake of completeness, the synthesis of Maruyama[136] about the role of elements is given in Table 2.4 where the role of the newly used alloy elements as W or Re, are also reported.

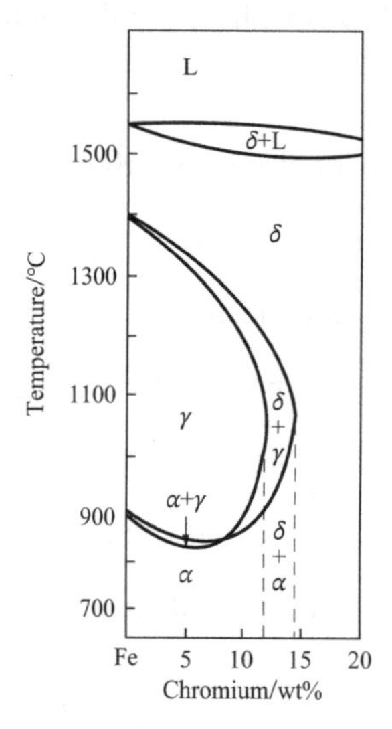

Figure 2.9 Range of liquid, austenite and ferrite (α and δ) phase in the iron-chromium constitution diagram with a carbon content below 0.01 wt%

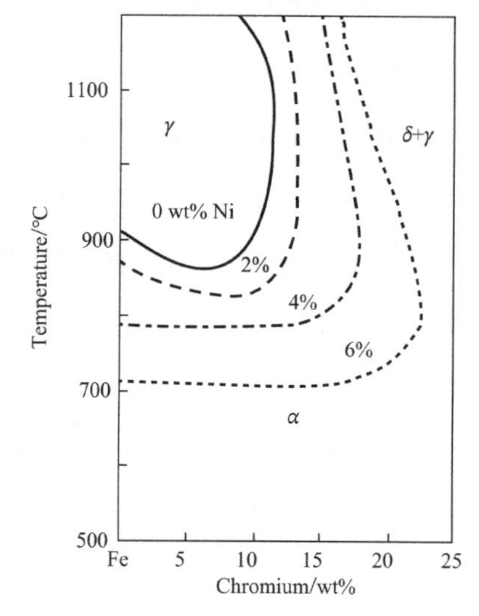

Figure 2.10 Influence of nickel on the range of the austenite phase field in the iron-chromium system[74]

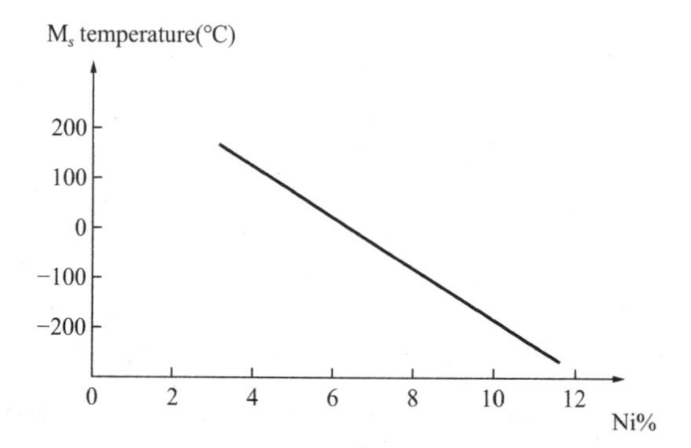

Figure 2.11 Martensite start temperature (M_s) plotted against nickel content for 18Cr wt%-0.04C wt% steel[112]

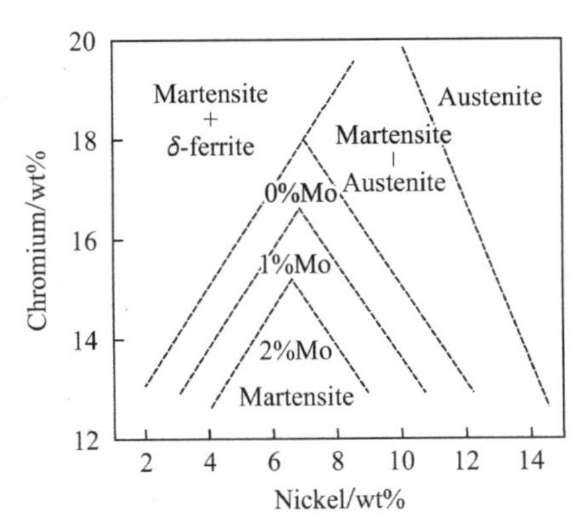

Figure 2.12 Experimental diagram showing the boundaries of the austenite, ferrite and martensite phases as a function of Cr, Ni and Mo concentration for 0.01 wt% C after austenitization at 1050 °C and air cooling[109]

Role of elements[136] Table 2.4

Element	Merit	Demerit
B	Improve creep strength and quench hardenability. Stabilize M23C6 particles and delay their coarsening.	Reduce impact toughness.
C	Necessary to make M23C6 and NbC.	
Co	Suppress δ-ferrite. Decrease D.	
Cr	Improve oxidation resistance. Lower Ms. Raise A1. Main element of M23C6.	Increase D.
Cu	Suppress δ-ferrite.	Promote precipitation of Fe_2M.
Mn		Increase D and reduce creep strength. Lower A1. Promote $M_{23}C_6$.
Mo	Lower Ms. Raise A1. Solid solution hardening.	Accelerate growth of $M_{23}C_6$.
N	Necessary to make VN.	
Nb	Form MX and contribute to strengthening.	Promote precipitation of z phase.
Ni		Increase D and reduce creep strength. Lower A1.
Re	Prevent the loss of creep rupture strength. Lower Ms.	Lower A1.
Si	Improve oxidation resistance.	Increase D and reduce creep strength.
V	Form MX and contribute to strengthening.	
W	Lower Ms. Raise A1. Delay coarsening of M23C6 particles. Solid solution hardening.	

Note: "D" designates the austenite grain size.

Carbon and some elements of addition have influence on Ms. An empirical formula as following can explain the details[109].

$$M_s(°C) = 540 - 497C - 6.3Mn - 63.3Ni - 10.8Cr - 46.6Mo(wt\%) \quad (2\text{-}1)$$

It can be deduced from the above equation that the Ms of material 15-5PH is 188.5 °C. The temperature predicated by the formula is found in good agreement with our experimental result.

Stress

Applied stress modifies the energy stored in material and the structure of the reticulum. The influence of stress on the transformations of phases is obvious. These couplings were studied in the Institut Polytechnique de Lorraine of France. Many work shows that the mechanisms of nucleation are most affected[50][51][78]. The applied stress is either hydrostatic or uniaxial. On the one hand, in terms of the hydrostatic pressure, only the extremely high value of about the hundred MPa can make their influence notable on the kinetics of transformation. On the other hand, the uniaxial pressure, even weak, can affect the transformation. Thus Gautier[78] shows that the times of beginning and end of transformation depend on the applied pressure in the case of an isothermal transformation in eutectoid steel. In the same way, Patel[150] studies the influence of stresses on the martensite start temperature. Hydrostatic pressure delays the transformation whereas compression or tensile stresses make the transformation to start earlier. However, in the case of the transformations with continuous cooling, if the cooling rate is relatively high (> 1°C/s), it seems that low uniaxial pressures have little influence on the kinetics of transformation.

2.3.4 Phase transformation models

Kinetic models

The first model of transformation of phase was proposed by Johnson and Mehl[102], then Avrami[10][11], in order to predict evolution of the proportions of pearlite.

$$z = 1 - \mathrm{e}^{-\frac{\pi}{3}NG^3t^4} \tag{2-2}$$

with

z : voluminal proportion of pearlite

N : rate of nucleation

G : rate of growth

t : time

These models suppose that the pearlite appears through nucleation then growth, depending on austenite. If the mechanisms are different, these models are used for the transformation of ferrite and bainite[12][205].

The equation of Johnson-Mehl-Avrami is written for isothermal transformation, whereas the requirements in calculation come mainly from the heat treatments, which are anisothermal. That's why some authors propose modifications of the preceding model to take into account the anisothermal effects. Thus we will note the models of Inoue[98], and then these models were further developed in INPL (Institut Polytechnique de Loraine)[65][88][172]. These models considered not only the effects of anisothermy but also the effects due to the applied stresses and the percentage of carbon.

The martensitic transformations are treated separately, because they are considered as independent of time. The empirical law of Koistinen and Marburger[108] gives the voluminal fraction of martensite according to the temperature. The theoretical justification of this equation was given by Magee[132]:

$$z_\alpha = z_\gamma\left(1 - e^{-\beta < M_s - T >}\right) \tag{2-3}$$

where z_α and z_γ are the voluminal proportions of martensite and austenite respectively; M_s is martensite start temperature; β is coefficient depending on material; T represents temperature.

Phenomenological models

The generalization of the models based on laws of the Johnson-Mehl-Avrami was done on phenomenological considerations and did not bring a really satisfactory application, because these models on the physical bases are difficult to use in practice. Their principal disadvantage comes from their specificity to describe single type of transformation. However in the case of the heat treatments or of welding, a same structure is prone to various transformations dictated mainly by the field of temperature. This is why one seeks for industrial applications and simple equations of evolution, which are applicable to varied solutions and easy to establish in a computer code. Other types of purely phenomenological models were developed.

Thus Leblond[114][115] proposed a model based on a law of simplified evolution utilizing a proportion of transformed phase in equilibrium and a constant of time:

$$\dot{z} = \frac{z_{eq}(T) - z}{\tau(T)} \tag{2-4}$$

with

z : voluminal proportion of new phase

$z_{eq}(T)$: voluminal proportion of phase in equilibrium

$\tau(T)$: constant of time

This model, identified from the CCT diagram, gives good results for the ferritic and perlitic transformations but it is not ready to reproduce the faster transformations correctly. Leblond introduces a dependence on dT/dt, which enables to make its model usable for the bainitic and martensitic transformations. However, the principal difficulty of use of the Leblond model lies in its identification.

Therefore, Waeckel[201][202] proposed another possibility, an easily identifiable model starting from the CCT diagram and able to reproduce thermal histories of welding. The purpose to limit the identification to the CCT diagram led Waeckel to limit the intervening variables in the transformations of phase. A law of evolution of this type is proposed:

$$\dot{\underline{z}} = f(T, \dot{T}, \underline{z}, d) \tag{2-5}$$

with

$\underline{z}$: proportion of considered phase

d : grain size of austenite

Contrary to the model of Leblond, in the model of Waeckel, $\dot{T}$ is an internal variable, which seems to rather close to the reality in this type of model owing to the fact that the kinetics is anisothermal. Also it is noted that the temperature M_s is not a metallurgical state, but as it can vary during transformation. It was blended in the metallurgical variables. Moreover the voluminal proportion of martensite is treated separately and follows the equation of Koistinen and Marburger[108]. The grain size of austenite is a parameter but not a variable. The function f is known in a certain number of states, and one can deduce the voluminal factions of the phases by linear interpolation. This model describes well the transformations of phases during cooling at a constant rate.

We will finish this review of the various models of phase transformation by work of Martinez, which was carried out at the CEA de Grenoble and from which we serve for calculations of thermal-metallurgical-mechanics. Martinez extends the model of Waeckel and adds the dependence on the grain size of austenite d and of the percentage of carbon c. Many tests were carried out to trace the CCT of a nuance of steel according to conditions of austenitization and carbon contents during cooling. Martinez proposes a law of evolution of this type:

$$\frac{\dot{z}_\gamma}{z_\gamma} = f(T, \dot{T}, z_\gamma, d, c) \tag{2-6}$$

where

c : carbon concentration

d : grain size of austenite

In this model, the variables c and d are calculated from the laws depending only on material and thermal. This model was established[21] in the calculation code of the CEA Cast3M[194]. Thermal-metallurgical procedure was suggested to calculate the quantities T, $\dot{T}$, z_γ, d and c, which are an iterative procedure in five stages:

1) T and $\dot{T}$ are calculated by a nonlinearly thermal calculation depending on the boundary conditions (conduction, convection, radiation), input sources and characteristics of material.

2) The carbon concentration is calculated by an equation of diffusion depending on $\dot{T}$ and the coefficient of activity of carbon.

3) The grain size d is determined by the equation of Grey Higgins depending on the temperature T.

4) The proportions of phases are calculated either by linear interpolations, or by an explicit equation in the case of the martensitic transformation. The result depends on T, $\dot{T}$, d and c.

5) Finally one calculates the latent heat of transformations.

In conclusion of this part, we propose a small assessment of these various models in Table 2.5.

Advantages and disadvantages of kinetic and phenomenological models Table 2.5

	Advantages	Disadvantages
Kinetic model	Simply use	Adapted to only one type of transformation
Phenomenological model	1. Exhaustive description of the various transformations. 2. Dependence on the carbon concentration and the size of grain.	1. Few physical concerns. 2. The validity of nonconstant $\dot{T}$ is hypothetical. 3. Calculation of interpolation is long.

2.4 Austenitic grain size and its influences

In welds, grain size is greatest at the fusion zone and then gradually gets smaller in the heat-affected zone (HAZ) until the base metal is reached. Therefore, grain size is important for study of mechanical behaviours (residual stress, damage ...). Calculation of the grain size can be useful for three reasons as following:

- ➢ This parameter is important in itself (it is involved in hot failure phenomena).
- ➢ Grain size can influence the transformations as mentioned above.
- ➢ Grain size has influences on mechanical behaviour of metal.

2.4.1 Grain size calculation model

Changes in grain size only concern the austenitic phase and depend on temperature changes and the proportion of this phase.

The evolution of grain size is calculated by using an extension of the conventional equation proposed by Alberry & Jones[6] and Ikawa et al.[100]:

$$\dot{G}^{a} = C\exp\left(-\frac{Q}{RT}\right) \tag{2-7}$$

where

G : grain size

T : absolute temperature

R : constant for gases

a, C, Q are positive constants. (Generally, $a = 4, C = 0.4948 \times 10^{14}$ mm^4/s, $Q/R = 63900$ gives satisfactory results.)

In addition, Ashby and Easterling have proposed a more general equation:

$$\dot{G} = f(G)\exp\left(-\frac{Q}{RT}\right) \tag{2-8}$$

This conventional equation is limited to cases in which the proportion of austenite is constant or decreasing. In fact, if the proportion of austenite increases, two phenomena can be observed physically: a) the size of the grains already existing increases; b) new grains are formed with an initial grain size of zero (or, at least, much lower than the existing size).

Generalization of the conventional equation was given by Leblond[114] as the following expressions:

$$\frac{\mathrm{d}}{\mathrm{d}t}(G^{a}) = \begin{cases} C\cdot\exp\left(-\dfrac{Q}{RT}\right) - \dfrac{\dot{z}}{z}G^{a} & \text{if} \quad \dot{z} > 0 \\ C\cdot\exp\left(-\dfrac{Q}{RT}\right) & \text{if} \quad \dot{z} \leqslant 0 \end{cases} \tag{2-9}$$

2.4.2 Effect on mechanical properties

Grain size has an important influence on mechanical behaviours. Generally, a small grain size in HAZ is beneficial to strength and toughness. The dependence of the yield strength on grain size is given by the Hall-Petch relationship[152]:

$$\sigma_y = \sigma_i + \frac{k_y}{\sqrt{d}} \tag{2-10}$$

where d is the grain diameter, σ_y is the yield stress, σ_i is the friction stress opposing the movement of dislocations in the grains and k_y is the fitting parameter (a material constant).

In fact, the Hall-Petch relationship or Hall-Petch Law is a relation in materials science that deals with the connection between the grain size, or crystallite size, and the yield point of a material. This relation says that the larger the grain size of a crystalline material, the weaker it is.

The derivation of the Hall-Petch equation relies on the formation of a dislocation pile-up at a grain boundary, one which is large enough to trigger dislocation activity in an adjacent grain. Yield in a polycrystalline material is in this context defined as the transfer of slip across grains. A larger grain is able to accommodate more dislocations in a pile-up, enabling a larger stress concentration at the boundary, thereby making it easier to promote slip in the nearby grain[107]. For martensitic structures, however, dislocation sources are found at grain boundaries, which are different from Hall-Petch approach that considers dislocation sources within individual grains. The increase in strength due to martensite lath size is given by:

$$\sigma_G = 115(\overline{L})^{-1} \tag{2-11}$$

where $\overline{L}$ is the mean linear intercept taken on random sections and at random angles to the length of any lath section.

In the case of phase transformation, in low carbon microalloyed steels, ferrite grain sizes and precipitation states are important factors, which affect the strength-toughness relationship. The ferrite grain size is a function of the austenite grain size after austenite recrystallization, the amount of retained strain in the austenite before the start of transformation, and the cooling rate through the transformation regime[156].

There is no general mechanism, by which grain refinement improves toughness. The argument for steels is that carbide particles are finer when the grain size is small. Fine particles are more difficult to crack and any resulting small cracks are difficult to propagate, thus leading to an improvement in toughness.

2.5 Measurement of phase proportions

This part introduces the measurement of the proportions of phase during a thermal cycle. There exist three possibilities to determine the proportions of phase. Resistivity as well as the magnetic permeability are very different for the γ and α phases, and the constitution of the two measurable quantities leads to the proportions of phases[90][92]. Here, we use the dilatometry to measure the proportions of phase. A test of dilatometry consists of heating and cooling of a sample, and measurement of its extension. Figure 2.13 gives an example of dilatometry. Proportions of phase can be deduced from this curve because of the difference between the dilation coefficients of γ phase and that of α phase.

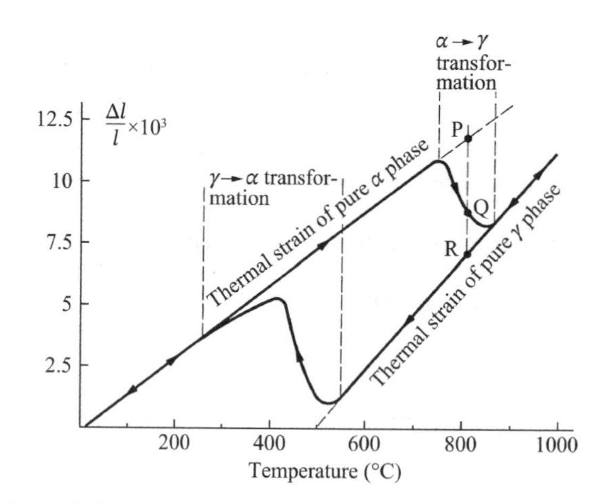

Figure 2.13 Test of dilatometry on A508 steel (heating: 30 °C/ s; cooling: −2 °C/s)[116]

Leblond shows, by considering two elastoplastic phases whose elastic characteristics are the same, which the partition of the total deformation is equal to:

$$E^{tot} = E^{e} + E^{thm} + E^{p} \tag{2-12}$$

$$E^{thm} = <\varepsilon^{e}>_{v} + <\varepsilon^{thm}>_{v} + <\varepsilon^{p}>_{v} \tag{2-13}$$

where, $<>_{v}$ represents the average volume.

$$E^{thm} = < \varepsilon^{thm} >_{v} = (1 - z_\gamma)\varepsilon_\alpha^{th} + z_\gamma\left(\varepsilon_\gamma^{th} + \Delta\varepsilon_{\alpha-\gamma}^{T_{ref}}\right) \tag{2-14}$$

The above equation can be written:

$$E^{thm} = [(1 - z_\gamma)\alpha_\alpha + z_\gamma\alpha_\gamma](T - T_{ref}) + z_\gamma\Delta\varepsilon_{\alpha,\gamma}^{T_{ref}} \tag{2-15}$$

$$z_\gamma(T) = \frac{E^{thm} - \alpha_\alpha(T - T_{ref})}{(\alpha_\gamma - \alpha_\alpha)(T - T_{ref}) + \Delta\varepsilon_{\alpha,\gamma}^{T_{ref}}} \tag{2-16}$$

We can calculate the volume proportion of austenite from the above equations.

2.6 Phase transformation induced plasticity (TRIP)

Transformation induced plasticity (TRIP), ε^{pt} in Figure 2.14, can be defined as the anomalous plastic strain observed when metallurgical transformation occurs under an external stress much lower than the yield limit. From a technological point of view, it plays an essential role in many problems, in particular for the understanding of residual stresses resulting from welding operations and for their prediction in practically significant cases. From an experimental point of view, it is usually characterized in a TRIP test where a constant stress is applied during transformation under prescribed cooling conditions.

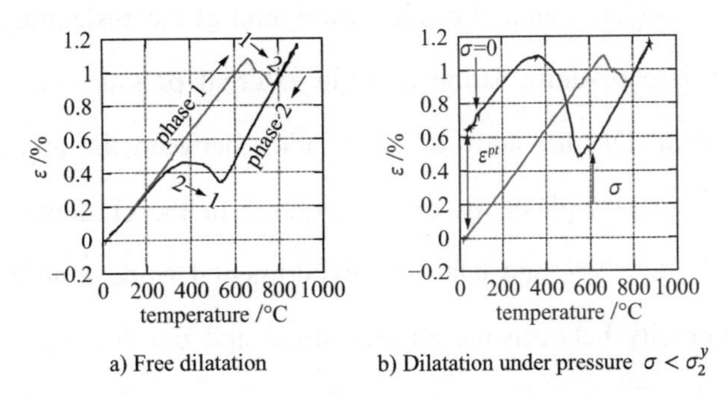

a) Free dilatation　　b) Dilatation under pressure $\sigma < \sigma_2^y$

Figure 2.14 Dilatation of uniaxial tests of phase transformation without/with applied stress[45]

2.6.1 TRIP mechanisms

From a phenomenological point of view, the transformation plasticity results from the coupling between an external stress and the evolution of proportions of phase involved. A transformation, which is accompanied by a variation of volume, causes a plastic strain of the softest phase. In general, several faces of transformation move in different directions, so that the plastic strains of transformation are not added from one to others and it appears a null macroscopic residual stress field. This is why, there is not remain of residual deformation at the end of a cycle of free dilation.

On the other hand, various work for a long time showed that the applied stress, even low, during the transformation, could lead to an additional deformation with the deformation of transformation, in the direction of the loading of pressure. It was called phase transformation induced plasticity (TRIP). Two mechanisms can explain this deformation of transformation.

Greenwood & Johnso's mechanism

The first model is the mechanism of Greenwood & Johnson. It expresses the fact that the transformation plasticity is induced by the applied pressure. This kind transformation plasticity arises from micro plasticity in the weaker austenitic phase induced by the difference of specific volume between the phases. This plastic flow is oriented by the external load. Many experimental results show that for low levels of stresses, the rate of deformation of transformation plasticity is proportional to the applied pressure, the variation of volume associated with the phase shift, the development of the transformation and inversely proportional to the yield stress of the softest phase.

For levels of stress higher than half of the elastic limit of the austenitic phase, the relation of linearity between the rate of deformation and the external pressure ceases. The deformation becomes more important. On the study of the relation between the applied pressure and the deformation of transformation plasticity, one can quote, in the chronological order, the work completed by[155][85][43][77][171][195], and lately the experimental work done by[28] and[153]. All this work confirms the linearity between the applied stress and the deformation of transformation plasticity as long as the pressure does not exceed a certain fraction of the elastic limit of the austenitic phase. This fraction depends on study of ferrous alloy.

Some experimental work[77][195][28][153] was dedicated to study the relation between the rate of deformation of transformation plasticity and the development of this transformation. To study this relation means to study the kinetics of the transformation plasticity: in other words, it is the study of the variation of the deformation according to the proportion of daughter phase induced by cooling. Experimental work shows a relation of non-linearity between these two quantities, the variation of the transformation plasticity being increasingly higher at the beginning of the transformation. This relation changes according to the metallurgical type of transformation during cooling.

It was verified that the mechanism initially proposed by Greenwood & Johnson is mainly applied to describe the deformation of transformation plasticity in the case of the ferritic transformation and, to a more extent, in the case of the bainitic transformation. Indeed in the case of the phase shifts without diffusion, i.e. the martensitic transformation, only this mechanism cannot be responsible for the phenomenon of transformation plasticity. Therefore the second mechanism proposed by Magee[133] is presented. It explains the orientation of the transformed zones by external stresses.

Magee's mechanism

Mechanism of Magee corresponds to the formation of selected martensitic variants resulting from the applied stress. Micrographic analysis shows that martensite has a crystallographic structure derived from that of austenite. Its final structure is obtained by a deformation of the initial phase, in particular a shearing. It has morphology of slats or plates and presents the relations of privileged orientation and the parent phase. When pressure is not applied, the auto-accommodating arrangement of the plates is like that the energy stored in material is minimized and the deformation measured at the end of the transformation is practically null. When the transformation occurs under pressure, the martensite plates are orientated by the external pressure. It is the phenomenon of transformation plasticity of Magee.

In the case of a monocrystal of parent phase transformed during cooling under constant and uniform external pressure, the deformation associated with the transformation has a limited value and does not depend on the applied stress[95]. This deformation is reversible when the transformation is reversed. However it is subject to the influence of the crystalline orientation.

For the polycrystal, the contribution of this mechanism depends on the distribution of orientation of the martensite plates. In general, the measured deformation in the direction of the stress is weaker than that obtained in the monocrystal. It is a function of the orientation of the products. There is no more linearity between this deformation and the development of the transformation, and it depends on the conditions of loading[77][195][68][69][15].

The relative importance of these two mechanisms depends on the material and the transformation under consideration. Strictly speaking, both mechanisms are generally present in diffusional and diffusionless transformations. However, general speaking, the preponderance of the mechanism of Magee is more than that of the mechanism of Greenwood & Johnson when the transformation is martensitic, depending on the difference in specific volume between the parent phases and daughter phases. The dilation of martensitic transformation is weak, and should not generate any transformation plasticity of the type of Greenwood & Johnson. It can be observed in thermoelastic alloys. For lowalloy ferritic steels, the dilation of transformation is driven and the mechanism of Greenwood & Johnson is dominating in stead of that of Magee.

2.6.2 TRIP models

Models based on Greenwood & Johnson's mechanism

Practically, all the models of transformation plasticity proposed and based on the mechanism of G&J are put in the form of a product of three functions:

$$\dot{\varepsilon}^{pt} = f_1(\Delta V/V; \sigma_\gamma^y) \cdot f_2(z; \dot{z}) \cdot f_3(S; \chi^{pt}) \tag{2-17}$$

with

$\Delta V/V$: voluminal variation associated with the transformation

σ_γ^y : yield stress of the mother phase

z : proportion of daughter phase transformed

S : deviatoric tensor of the applied stress

χ^{pt} : tensor of the internal stresses associated with plasticity with transformation

The function f_1 introduces the dependence of the rate of transformation plasticity with respect to the yield stress of the mother phase and the change of volume induced by transformation. The second function f_2 expresses the dependence according to the rate of development of transformation. Finally the last function f_3 introduces a relation between the applied pressure, the internal stress and the rate of transformation plasticity.

In the absence of internal stress and if the load applied during the transformation is constant and weaker than the elastic limit of the mother phase, the preceding relation is integrable and can then be put in the following form:

$$\varepsilon^{pt} = g_1(\Delta V/V; \sigma_y^\gamma) \cdot g_2(z) \cdot g_3(\sigma) \tag{2-18}$$

After a short recall of the historically unidimensional models, in particular developed by G&J, Abrassart and Desalos, we will present the principal multiaxial models.

- ***Greenwood & Johnson***[85]

Its expression is given as following:

$$g_1 = \frac{5}{6\sigma_y^\gamma}\frac{\Delta V}{V}; g_2 = 1; g_3 = \sigma \tag{2-19}$$

This uniaxial model can only envisage the finial value of the deformation of transformation plasticity for a constant load applied (weak) during phase changes.

- ***Abrassart***[4]

$$g_1 = \frac{3}{4\sigma_y^\gamma}\frac{\Delta V}{V}; g_2 = \left(3z - 2z^{3/2}\right); g_3 = \sigma \tag{2-20}$$

The introduction of a function g_2, depending on the proportion of daughter phase, authorizes the simulation of the deformation of transformation plasticity during phase transformation.

- ***Desalos***[54]

$$g_1 = k; g_2 = z \cdot (2 - z); g_3 = \sigma \tag{2-21}$$

This relation uses a constant coefficient k. The function g_2 is different from that of the preceding relation.

The expressions above developed to describe the evolution of the deformation of transformation plasticity present important gaps. These models cannot deal with the multiaxial problems and their application is not possible for low levels of stresses. Thus, we introduce the following multiaxial models.

- ***Leblond***

Leblond established a multiaxial and incremental formulation of the deformation of transformation plasticity. The modeling retained for this phenomenon is founded on theoretical and numerical studies of the mechanism of Greenwood and Johnson, where the mechanism of Magee is neglected. The mathematical model simulates the phenomenon of transformation plasticity in the case of ideally plastic phases[116][118]. The extension of this model to the case of materials to isotropic or kinematic work hardening is presented in reference[119].

Leblond supposes that the daughter phase is a spherical inclusion which grows inside an austenitic sphere. It considers the behaviour of the mother phase as plastic perfect whereas the daughter phase remains elastic. It obtains finally the expression below. The following formalism takes into account the isotropic work hardening of the phases. Rate of deformation of transformation plasticity is:

$$\dot{E}^{pt} = -\frac{3\Delta\varepsilon_{\gamma\alpha}^{T_{ref}}}{\sigma_\gamma^y\left(E_\gamma^{eff}\right)} \cdot S \cdot h\left(\frac{\Sigma^{eq}}{\Sigma^y}\right) \cdot (\ln z_\gamma) \cdot \dot{z}_\gamma \quad \forall z_\gamma \subset [0.03 \quad 1] \tag{2-22}$$

$\Sigma^{eq} = \left(\frac{3}{2}S_{ij} \cdot S_{ij}\right)^{\frac{1}{2}}$ macroscopic equivalent stress (von Mises)

$\dot{E}^{eq} = \left(\frac{2}{3}\dot{E}_{ij}^p \cdot \dot{E}_{ij}^p\right)^{\frac{1}{2}}$ rate of macroscopic equivalent plastic strain

$\Sigma^y = [1 - f(z)] \cdot \sigma_\gamma^y\left(E_\gamma^{eff}\right) + f(z) \cdot \sigma_\alpha^y\left(E_\alpha^{eff}\right)$ homogenized macroscopic ultimate stress

with

$f(z)$: correctional function resulting from digital simulations (see Table 2.6)

σ_γ^y : elastic limit of the phase γ

σ_α^y : elastic limit of the phases α

E : elasticity modulus, considered common to all the phases

S : homogenized macroscopic deviatoric tensor

E_γ^{eff} : macroscopic effective internal variable of isotropic hardening of the mother phase

E_α^{eff} : macroscopic effective internal variable of isotropic hardening of the daughter phase

$\Delta\varepsilon_{\gamma\alpha}^{T_{ref}}$: difference of compactness of phases α compared to phase γ at T_{ref}

$h\left(\frac{\Sigma^{eq}}{\Sigma^y}\right)$: term which translates the non-linearity of the transformation plasticity according to the equivalent macroscopic stress applied.

Leblond distinguished the condition of the low stresses from that of the high stresses. The rate of deformation of transformation plasticity presented above does not intervene if the homogenized macroscopic equivalent stress is strictly lower than the ultimate stress of the mixture of multiphase containing austenite. This ultimate stress is a function of the elastic limits of each metallurgical component which depends on the effective internal variables of hardening. A function h is also introduced in the formulation of the rate of deformation of transformation plasticity.

$$h\left(\frac{\Sigma^{eq}}{\Sigma^y}\right) = \begin{cases} 1 & \text{if } \frac{\Sigma^{eq}}{\Sigma^y} \leqslant 0.5 \\ 1 + 3.5\left(\frac{\Sigma^{eq}}{\Sigma^y} - \frac{1}{2}\right) & \text{if } \frac{\Sigma^{eq}}{\Sigma^y} > 0.5 \end{cases}$$

It introduced the nonlinearity of the rate of deformation with the level of the applied pressure. This function results from numerically micromechanical tests of transformation plasticity[118][55].

Note: To circumvent an awkward singularity when $z \to 0$, Leblond considers that the rate of deformation of transformation plasticity is null as long as the fraction of daughter phase is less than 3%.

Models based on Magee's mechanism

This very complex mechanism is often neglected even for the martensitic transformation, except where it is always dominating. Fischer starts with the principle which the transformation of plasticity results from the two mechanisms. It uses a micromechanically analytical approach for its modelling. It introduces a spatial arrangement of the martensite variants which it describes by the Euler angles. A function of distribution G takes into account the effect of orientation of the martensite variants.

By a rather complex mathematical process, based on a random distribution of the variants, it establishes the following formulation[69]:

$$\varepsilon^{pt} = \frac{5}{\sigma_y^*}\left[\left(\frac{\Delta V}{V}\right)^2 + \frac{3}{4}\gamma^2\right]^{1/2} \cdot S \tag{2-23}$$

with

$\sigma_y^* = \sigma_y^\alpha \left(\frac{1-\sigma_y^\gamma/\sigma_y^\alpha}{\ln(\sigma_y^\alpha/\sigma_y^\gamma)}\right)$: Average elastic limit of γ-α phase mixture

γ : Constant of shearing induced by the deformation of transformation

One can also quote another model of this phenomenon, the micromechanical models developed by Cherkaoui & Berveiller[34].

2.7 Multiphase mechanics and models

The simulation of the problems involved in the heat treatments or welding, like presented in the introduction, should take into account thermal, metallurgical and mechanical phenomena, more or less coupled. Thus, this part lists various equations to be solved as well as the type of model used. At last, we will present two classes of multiphase mechanical models in detail.

2.7.1 Formulation of the problem

Thermal problem

The thermal problem (Figure 2.15) which one seeks to solve follows the Furrier's law. In multiphase materials, latent heats of transformation are introduced into the equation of heat. In addition, structures can exchange heat by convection or radiation. The problem thus consists of determining the temperature field T and the heat flow $\underline{q}$, such as:

- ***Boundary conditions***

$$T = Td \quad \text{on} \quad \partial_1\Omega \tag{2-24}$$

$$q\underline{n} = \underline{qd} \quad \text{on} \quad \partial_2\Omega \tag{2-25}$$

$$\text{Convection} \quad q = h_c(T - T_{ext}) \tag{2-26}$$

$$\text{Radiation} \quad q = h_r(T^4 - T_{ext}^4) \tag{2-27}$$

- ***Heat equation***

$$\rho c\dot{T} = -div\underline{q} - \rho L\dot{z} + \rho r \tag{2-28}$$

- ***Constitutive relation***

$$\underline{q} = -k\overrightarrow{grad}(T) \tag{2-29}$$

Where q is the heat flux per unit; r is the heat flux per unit volume generated within the body; ρ is the mass density of material; c presents the specific heat capacity; k is the coefficient of thermal conductivity. L represents the latent heat of phase transformation.

Figure 2.15 Thermal boundary conditions Figure 2.16 Mechanical boundary conditions

Metallurgical problem

The metallurgical problem is strongly coupled with the thermal problem, which determines the evolution of the voluminal proportions of phases. The formulation takes the models of the type Waeckael or Martinez, which was introduced in Chapter 2.3.4.

$$\dot{z}_i = f(T, \dot{T}, z_i, d, c) \quad \text{and} \quad \sum_1^n z_i = 1 \tag{2-30}$$

Mechanical problem

The mechanical problem (see Figure 2.16) is more complex than thermal problem and our efforts will focus on this problem of multiphase.

- ***Boundary conditions***

$$\underline{U} = \underline{Ud} \quad \text{on} \quad \partial_3\Omega \tag{2-31}$$

$$\underline{F} = \sum \underline{n} \quad \text{on} \quad \partial_4\Omega \tag{2-32}$$

- ***Equilibrium***

$$\underline{div\sum(M)} = \underline{0} \qquad \forall M \in \Omega \tag{2-33}$$

- ***Constitutive relation***

It is to be determined.

2.7.2 Mechanical models of multiphase

There are many models which we will not present here. The following theses and books[56][69][153][135] propose a broadly bibliographical review of these models. We will introduce some particular models here. Philosophy adopted to deal with the mechanical problem is in order to obtain an explicit relation of the mixture behaviour, based on the behaviour of each phase. Thus, two methods are exploited to obtain the macroscopic behaviour of material. The first method uses the thermodynamics of the irreversible processes[80]. The second method aims to determine macroscopic laws through micro-macro method.

Models based on mixture of energy

- ***Inoue & Wang***[98][99]

The model of Inoue is one of the simplest mechanical models of multiphase. This model is elaborated based on linear laws of mixture of internal energies and the potentials of dissipation[124]. Inoue does not separate strain of transformation plasticity from the total train. The model is developed in plasticity or in viscoplasticity and the mechanical properties depend on the proportions of phases.

The partition of the strains is as follows:

$$\dot{E}^e = \dot{E} - \dot{E}^i \tag{2-34}$$

The recoverable and irreversible strains are written:

$$E^e = \left(\sum_{i=1}^{n} \frac{1+\upsilon_i}{E_i} z_i\right)\sum - \left(\sum_{i=1}^{n} \frac{\upsilon_i}{E_i} z_i\right) Tr(\sum)II + \left(\int_{t_0}^{t_1} \left[\sum_{i=1}^{n} \alpha_i z_i\right] \mathrm{d}t + \sum_{i=1}^{n} \beta_i (z_i - z_{io})\right) II \tag{2-35}$$

$$\dot{E}^{i} = \frac{1}{2\mu}\left\langle 1 - \frac{K(T,\kappa,z_i)}{J_2(S-X)}\right\rangle (S-X) \tag{2-36}$$

with

β_i : rate of variation of volume due to transformation i

μ : viscosity of homogeneous material

K : variable of hardening

κ, X : isotropic and kinematic hardening

- INPL

The INPL (Institut Polytechnique de Loraine) proposed many models. These models cover a broad field since they supposed plastic[50][88][172] or viscoplastic[7][8][43][78][126] behaviours, which can be coupled with the metallurgy[52] and the concentration of carbon[52][78] or grain size of austenite[126]. The behaviour of material is described by aggregate variables and the parameters of the behaviour law are calculated by linear mixtures of the characteristics of the phases[7][43][50]. Another type of model was developed where one considers a law of evolution of the plastic flow particular for each phase. It is supposed, in these models, that the plastic strain is the same for all the phases[126][78][52].

We present here the principal equations of the viscoplastic model[7]. The partition of the strains can be written as:

$$\dot{E} = \dot{E}^{e} + \dot{E}^{vp} + \dot{E}^{th} + \dot{E}^{met} + \dot{E}^{pt} \tag{2-37}$$

where metallurgical dilation is given by:

$$\dot{E}^{met} = \sum_{i=1}^{n} \varepsilon_{i0}^{tr} z_i II \tag{2-38}$$

the transformation plasticity by:

$$\dot{E}^{pt} = f(z_\gamma)\dot{z}_\gamma S \tag{2-39}$$

the viscoplastic strain by:

$$\dot{E}^{vp} = \left\langle \frac{J_2(S-X)-\kappa}{K}\right\rangle^{n} \frac{S-X}{J_2(S-X)} \tag{2-40}$$

- Hamata[89][90]

As what we presented previously, Hamata proposes a biphasic model for cast iron GS. Here also, the viscoplastic flow of each phase is dictated by its own threshold function. Moreover, Hamata identifies a viscotransformation plasticity:

$$\dot{E}^{pt} = \frac{3}{2}\left(\frac{\sigma_{eq}}{\lambda}\right)^n \frac{S}{\sigma_{eq}}\dot{z} \tag{2-41}$$

- ***Videaul*[195][196]**

In its study, Videau is interested in the interaction between classical plasticity and transformation plasticity. Its model considers a "classical" law of viscoplastic flow and a law of viscoplastic flow related to the transformation plasticity. Lastly, the various parameters kinematics are bounded by a matrix of interaction.

The macroscopic behaviour is generally obtained through a linear mixture on the characteristics of the phases. Then, their phenomenological approach is constrained to an experimental identification for all the proportions of phases and all the temperatures.

Models based on micro-macro method

- ***Leblond*[116][117][118][119]**

Leblond undertakes a theoretical study of the problem, from which it brings new results, whether the behaviour of the phases is either perfect elastoplastic or kinematic elastoplastic. The object of its study is A508 steel. By supposing that all the phases have the same characteristics thermoelastic, Leblond shows that the macroscopic rates of deformations can be put in the following form:

$$\dot{E} = \dot{E}^e + \dot{E}^{thm} + \dot{E}^p \tag{2-42}$$

The rates of elastic strain $\dot{E}^e$ and thermal $\dot{E}^{thm}$ usually follow linear mixture law. On the other hand, the rate of plastic strain is the sum of three terms proportional to $\dot{S}$, $\dot{T}$ and $\dot{z}$. The transformation plasticity appears here naturally. In Leblond model, the classical plastic strain rate $\dot{E}^{cp}$ is a sum of two terms: the first one, $\dot{E}^{cp}_{\Sigma}$, proportional to the stress rate, represents classical plasticity at constant temperature; the second one, $\dot{E}^{cp}_{T}$, proportional to the temperature, represents classical plasticity at constant applied stress.

If $\Sigma^{eq} < \Sigma^{y}$

$$\dot{E}^p = \dot{E}^{cp}_{\Sigma} + \dot{E}^{cp}_{T} + \dot{E}^{tp} \tag{2-43}$$

$$\dot{E}^{cp}_{\Sigma} = \frac{3(1 - z_\gamma)}{\sigma^y_\gamma} \frac{g(z_\gamma)}{z_\gamma E} S\dot{\Sigma}^{eq} \tag{2-44}$$

$$\dot{E}^{cp}_{T} = \frac{3(\alpha_\gamma - \alpha_\alpha)}{\sigma^y_\gamma} z_\gamma \ln(z_\gamma) S\dot{T} \tag{2-45}$$

and the formula of plasticity of phase transformation $\dot{E}^{tp}$ was given in Chapter 2.6.2 (see

Equation 2-17).

If $\Sigma^{eq} < \Sigma^{y}$

$$\dot{E}^{p} = \dot{\Lambda} S \tag{2-46}$$

where $\dot{\Lambda}$ is unspecified.

However, certain results are difficult to obtain analytically without many assumptions. This is why, calculation by finite elements (implemented in Sysweld) is necessary to know the yield stress of the mixture. Leblond then proposes the function f to compute the law of mixture (Table 2.6).

Values of functions $f(z)$ and $g(z)$ in Leblond model[117] Table 2.6

z	0	0.125	0.25	0.50	0.75	1.0
$f(z)$	0	0.044	0.124	0.391	0.668	1.0
$g(z)$	0	2.53	4.0	2.76	1.33	1.0

Leblond continues its study by introducing hardening on the level of the behaviour of each phase. The case of kinematic hardening is treated as that of isotropic hardening. Hardening has an influence on the transformation plasticity because it modifies the yield stress of the phases (Equation 2-17). This model can also give an account of an effect of memory of hardening between mother phases and daughter phases.

Leblond model including isotropic hardening:

If $\Sigma^{eq} < \Sigma^{y}$

$$\dot{E}_{\Sigma}^{cp} = \frac{3(1-z_{\gamma})\,g(z_{\gamma})}{\sigma_{\gamma}^{y}\left(E_{\gamma}^{eff}\right)\,z_{\gamma}E} S\dot{\Sigma}^{eq} \tag{2-47}$$

$$\dot{E}_{T}^{cp} = \frac{3(\alpha_{\gamma}-\alpha_{\alpha})}{\sigma_{\gamma}^{y}\left(E_{\gamma}^{eff}\right)} z_{\gamma}\ln(z_{\gamma})\,S\dot{T} \tag{2-48}$$

$$\dot{E}_{\gamma}^{eff} = \frac{2\Delta\varepsilon_{\alpha-\gamma}^{T_{ref}}}{1-z_{\gamma}}\ln(z_{\gamma})\,h\left(\frac{\Sigma^{eq}}{\Sigma^{y}}\right)\dot{z}_{\gamma} + \frac{g(z_{\gamma})}{E}\dot{\Sigma}^{eq} + \frac{2(\alpha_{\gamma}-\alpha_{\alpha})}{1-z_{\gamma}}\ln(z_{\gamma})\,\dot{T} \tag{2-49}$$

$$\dot{E}_{\alpha}^{eff} = -\frac{\dot{z}_{\gamma}}{z_{\gamma}}E_{\alpha}^{eff} + \theta\frac{\dot{z}_{\gamma}}{z_{\gamma}}E_{\gamma}^{eff} \tag{2-50}$$

If $\Sigma^{eq} < \Sigma^{y}$

$$\dot{E}^{p} = \frac{3}{2}\frac{\dot{E}^{eq}}{\Sigma^{eq}} S \tag{2-51}$$

$$\dot{E}_{\gamma}^{eff} = \dot{E}^{eq} \tag{2-52}$$

$$\dot{E}_{\alpha}^{eff} = \dot{E}^{eq} - \frac{\dot{z}_{\gamma}}{z_{\gamma}} E_{\alpha}^{eff} + \theta \frac{\dot{z}_{\gamma}}{z_{\gamma}} E_{\gamma}^{eff} \tag{2-53}$$

- ***Coret & Combescure***[45][46]

The model of Coret & Combescure is much more numerically oriented, and ignores an a priori constitutive law for each phase. This model is developed on the basis of four assumptions:

➢ The rate of total macroscopic strain consists of the rate of microscopic strain of the phases and the transformation plasticity.

➢ They uncouple classical plasticity from the transformation plasticity. That means neglecting the interactions between these two parts.

➢ The local and macroscopic mechanical magnitudes are connected via the assumption of Taylor. In other words, the rate of homogenized macroscopic strain withdrawn from the rate of strain of transformation plasticity is equal to the microscopic rate of strain of each phase.

➢ The homogenized macroscopic stress is obtained by a linear law of mixture balanced by the volume fraction of the phases. That amounts to neglecting nonlinearity (according to the proportion of mother and daughter phase) of the mechanical behaviour of a mixture of phases.

The strain rate equations are

$$\dot{E}^{tot} = \dot{E}^{cp} + \dot{E}^{pt} \tag{2-54}$$

$$\dot{E}^{cp} = \dot{\varepsilon}_i \tag{2-55}$$

$$\dot{\varepsilon}_i = \dot{\varepsilon}_i^{e} + \dot{\varepsilon}_i^{thm} + \dot{\varepsilon}_i^{vp} \tag{2-56}$$

The stress equation is

$$\Sigma = \sum_{i=1}^{n} z_i \underline{\sigma_i} \tag{2-57}$$

- ***Others***

1. Taleb and Sidoroff[182][153]

The authors extend the study of Leblond and modify some assumptions. They do not neglect the elastic strain of austenite and do not consider it only plastic, whatever the proportion of formed phase. They arrive finally at an expression very close to that of Leblond.

2. Fischer[69][71]

Fischer adopts a micromechanical analysis, including not only the mechanism of Greenwood and Johnson but also the mechanism of Magee. Its very thorough work leads to the following expression:

$$\dot{E}^{pt} = \frac{5}{4\sigma_\gamma^*}\left[\left(\frac{\Delta V}{V}\right)^2 + \frac{3}{4}\gamma^2\right]^{\frac{1}{2}} S \tag{2-58}$$

where σ_γ^* is calculated from the yield stresses of the elastic phase and the ferritic phases. γ represents the deviatoric part of the transformation.

3. Diani[56][58]

This study is to treat the transformation plasticity in the case of fast loadings. Diani undertakes a theoretical study based on work of Stringfellow[175] concerning memory-shape alloys. Its main contribution relates to the modelling of the coupling between applied pressure and the evolution of the martensitic transformation.

2.8 Summary

We introduce the material studied, 15-5PH, a martensitic precipitation hardening stainless steel, which undergoes phase transformation during heating and cooling. Our numerical model to be developed is based on the study of this martensitic steel. Hence, there are detailed introductions of the material structure and the behaviours of phase transformation in this chapter.

It is shown that the comprehension and the mechanical modelling of the problems with phase change are particularly complex. The phase transformation must be imperatively taken into account for the correct simulation of the mechanical problem considered. Therefore, we present the concepts and mechanisms of phase transformation and transformation induced plasticity. And then the various models of phase transformation and transformation induced plasticity are listed and explained to widen and further the understanding of the achievement and development of this field. Several particular models will be used and studied in our succeeding work of numerical simulation.

In addition, we also presented various mechanical models of multiphase built either on an energy law of mixture, or on a microphone-macro approach. The common point of these two approaches is to explicitly seek the law of behaviour of multiphase material.

Chapter 3 Damage Mechanics and Welding Damage

3.1 Introduction

Damage and cracking deteriorate the cohesion because of voluminal or superficial discontinuities induced by welding in materials. These damage phenomena are associated with the decrease in the material properties due to the nucleation and growth of microvoids and microcracks. From the physical point of view, the damage or deterioration of materials is an irreversible and progressive physical process. The entropy is increasing during such a damaging process. Depending on the nature of the material, the type of loading, and the temperature, the damage appears in various ways, including brittle damage, ductile damage, creep damage, low cycle fatigue damage and high cycle fatigue. The cause of these material imperfections could be due to several processes:

➢ The creep damage due to the viscoplastic strain, which leads to intergranular decohesions.

➢ The high cycle fatigue damage where the local microplasticity occurs.

➢ The fragile cleavage damage.

➢ The ductile damage for which large plastic strains induce dislocation stacking and micro cavities.

In terms of different levels of size studied, damages mentioned above are classified to three categories: microscale, mesoscale, and macroscale.

➢ At the microscale, damage is caused by the accumulated deterioration due to microstresses in the neighborhood of defects, interfaces and by breaking of bonds. When the size of these cracks reaches dimensions of the order of the micron, they could be regarded as microscopic cracks.

➢ At the mesoscale, i.e. the level of the so-called representative volume element (RVE), damage is associated with the growth of microcracks or microvoids that initiate the final macroscopic crack. The size of the RVE for metals is 0.1 mm^3.

➢ At the macroscale, it is the growth of this macroscopic crack.

In terms of microscale and mesoscale, we can study them by means of damage variables and mechanics of continuous media defined at the mesoscale level. The third stage is usually studied using fracture mechanics with variables defined at the macroscale level.

The micro or meso mechanical approach is much more physically orientated, and it investigates microscopic behaviours and elementary mechanisms (grain, system of slip and dislocation) of metallic materials. The smaller the scale is, the more anisotropic unit studied is. Based on the analysis of internal variables at microscale, the macromechanical behaviour can be deduced by using the homogenization method. Although the micro approach seems to be more realistic, it is less likely to be modelled successfully because microscopic material parameters are difficult to determine and the considerable number of equations lead to extremely long computing times.

Another modelling approach is phenomenologic. It is founded on the introduction of state variables associated with the various phenomena, which are revealed by the experiments. These phenomena are described within the framework of thermodynamics of irreversible processes[80]. Some models have been developed by studying cavities or porous solid plasticity[160][87][163]. This kind of approach is based on a growth rate of the cavities inside a matrix with the elastoplastic behavior[160]. This theory assumes the isotropy of damage and uses a scalar variable to describe the voluminal fraction of the cavities. Others are based on continuum damage mechanics (CDM)[105][158][29][124]. The theory assumes that damage is one of the internal variables of constitutive equations that govern the irreversible processes in the material.

In our study, the ductile damage is focused on metallic materials. The literature, current situation and various models of ductile damage will be presented in this chapter. Damage behaviour in single phase will be emphasized while the behaviour of ductile damage for multiphase materials will be involved in next chapter.

3.2 Phenomenological aspects

The difficulty of defining a mechanical damage variable lies in the fact that there is little difference between a damaged volume element and a virgin one in macroscopic scale. Therefore, it is necessary to use internal variables which are representative of the deteriorated state of the matter. Following are some types of damage measurement envisaged[124]:

➢ Measurements at the scale of microstructure (density of microcracks or cavities) that lead to microscopic models. These can be integrated into macroscopic behaviour via homogenization technique.

➢ Global physical measurements (density, resistivity etc.) that can be converted into the mechanical properties of material by using the proper model.

➢ Global mechanical measurements (modification of elastic or plastic properties) that are easier to interpret in terms of damage variable via using the concept of effective stress introduced by Rabotnov[158].

3.2.1 Damage variable

Representative volume element

Continuum mechanics deals with quantities defined for the mathematical point. From a physical point of view, this means some kind of homogenization in a certain volume. This Representative Volume Element (RVE) must be small enough to avoid smoothing of high gradients; meanwhile they have to be large enough to represent an average of the microprocesses. Certainly the size of the RVE depends on the chosen material, e.g. 0.1 mm^3 for metals and ceramics, 1 mm^3 for polymers and composites, and 10 mm^3 for wood[124].

Definition of damage variable

The damage may be interpreted at the microscale as the creation of microsurfaces of discontinuities: breaking of atomic bonds and plastic enlargement of microcavities. At the mesoscale, the number of broken bonds or the pattern of microcavities may be approximated in any plane by the area of the intersections of all the flaws with that plane. In order to manipulate a dimensionless quantity, this area is scaled by the representative volume element. This size is very important in the definition of a variable continuous in terms of continuum mechanics.

The damage variable, as introduced by Kachanov[105], was defined on the basis of the irreversible processes leading to nucleation and growth of microvoids and microcracks in the entire volume of the sample. All types of voids and cracks (inter- and transgranular) that deteriorate the integrity of the material are accounted for.

Consider a damaged part with a RVE in one point orientated by a normal $\vec{n}$. δS is the area of intersection of the considered plane with the representative volume element (RVE). δS_D is the effective area of intersections of all microcracks and microcavities in δS (Figure 3.1). The damage value $D(\vec{n})$ attached to this point is defined as:

$$D(\vec{n}) = \frac{\delta S_D}{\delta S} \tag{3-1}$$

In the general case of anisotropic damage consisting of cracks and cavities with preferred orientations, the scalar variable depends on the orientation of the normal.

If microcracks and cavities are uniformly distributed, the damage variable will not depend on the orientation of $\vec{n}$. For homogeneous damage, damage variable could be simplified, and onedimensional variable is defined as:

$$D = \frac{S_D}{S} \tag{3-2}$$

This is the case of isotropic damage. We will restrict our study to the case of isotropic damage for most cases. From the definition of damage variable, the value of the scalar variable D is limited as following:

$$0 \leqslant D \leqslant 1 \tag{3-3}$$

When $D = 0$, it means that the material in the RVE is undamaged, and we may call such material as "virgin" material. $D = 1$ means that the RVE material is broken into two parts.

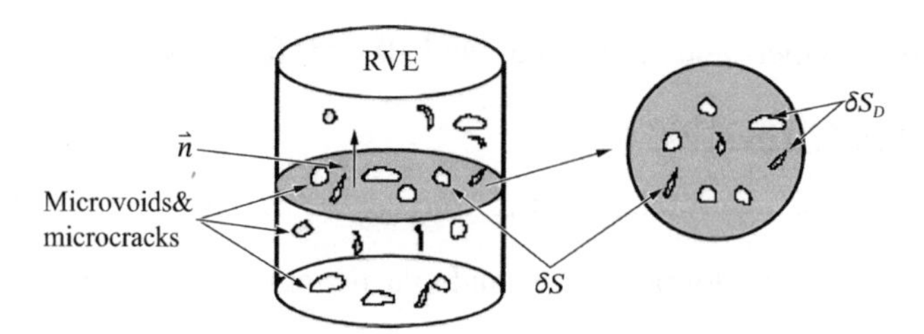

Figure 3.1　Definition of the damage variable

3.2.2 Effective stress

The introduced damage variable leads directly to the concept of effective stress. In the case of uniaxial tension, if all the defects are open, the cross section of isotropic damaged material is no longer S but $S - S_D$. Thus the effective stress becomes:

$$\tilde{\sigma} = \frac{F}{S - S_D} \tag{3-4}$$

It will be therefore written in terms of the effective stress as:

$$\tilde{\sigma} = \frac{F}{S\left(1 - \frac{S_D}{S}\right)} = \frac{\sigma}{1 - D} \tag{3-5}$$

Evidently $\tilde{\sigma} \geqslant \sigma$, $\tilde{\sigma} = \sigma$ for a virgin material and $\tilde{\sigma} \to \infty$ at the moment of facture.

3.2.3 Strain equivalence principle

We intend to avoid the micromechanical analysis for each type of defect and damage mechanism. Therefore, a themodynamical description is used in a mesoscale postulating the following principle. Lemaitre proposed the principle of strain equivalence[124]:

Any deformation behaviour, whether uniaxial or multiaxial, of a damaged material is represented by the constitutive laws of the virgin material in which the usual stress is replaced by the effective stress.

The uniaxial linear elastic law of a damaged material is written as:

$$\varepsilon_e = \frac{\tilde{\sigma}}{E} = \frac{\sigma}{(1 - D)E} \tag{3-6}$$

where E is Young's modulus.

3.2.4 Damage measurement

Damage is not easy to measure directly. Its quantitative evaluation, like any physical value, is linked to the definition of the variable chosen to represent the phenomenon. We can measure damage by measuring the modification of the mechanical properties of elasticity, plasticity and viscoplasticity.

The method of variation of the modulus of elasticity, which is a non-direct measurement, is preferred because of its simplicity (Figure 3.2).

Recall effective stress in the case of uniaxial tension, which we mentioned in the previous section as:

$$\tilde{\sigma} = \frac{\sigma}{1-D} = E\varepsilon_e \tag{3-7}$$

$E(1-D) = \tilde{E}$ could be interpreted as the elastic modulus of the damaged material. Damage variable D could be written as:

$$D = 1 - \tilde{E}/E \tag{3-8}$$

where E is Young's modulus of undamaged material. $\tilde{E}$ is Young's modulus of damaged material.

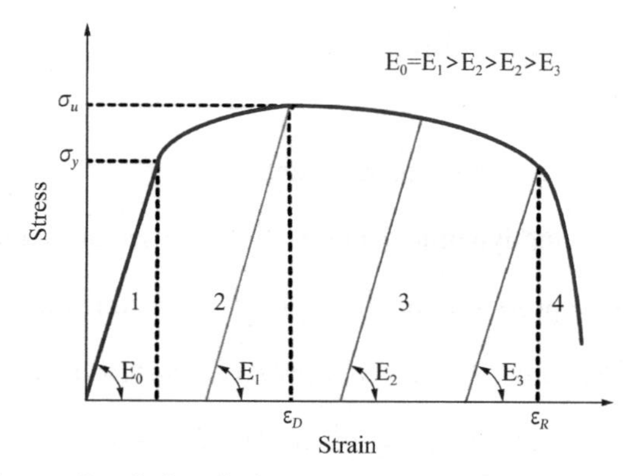

Note: Zone 1: elastic stage; Zone 2: elastoplastic stage; Zone 3: elastoplastic stage in which damage occurs; Zone 4: initiation and propagation of cracks leading to final rupture.

Figure 3.2 Schematic diagram of stress vs. strain and variablation of Yound's modulus

Although the principle is very simple, this measurement is rather tricky to perform for the following reasons:

➢ The measurement of modulus of elasticity requires precise measurement of very small strain.

➢ Damage is usually localized, thus requires a very small base on the order of 0.5-5mm.

➢ The best straight line in the strain-stress graph representing elastic loading or unloading is difficult to define.

➢ At high temperature, metal becomes weak and strain-stress curve indicates strongly non-linear.

Besides the very useful method of variation of the Young's modulus, thanks to the development of technology, there are some other approaches for the choice to measure damage[125]:

➢ Ultrasonic wave propagation: a technique based on the variation of the elasticity modulus consists in measuring the speed of ultrasonic waves.

➢ Variation of the microhardness: based on the influence of damage on the plasticity-yield criterion through the kinetic coupling.

➢ Variation of density: based on the measurement of decrease of density with apparatuses of Archimedean principle.

➢ Variation of electrical resistance: the effective intensity of the electrical current can be defined in the same way as the definition of effective stress.

➢ Acoustic emission: a good method to detect the location of the damage zone, but the results remain qualitative as far as the values of variable D are concerned.

3.3 Thermodynamics of isotropic damage

In the framework of thermdynamics, two potentials are introduced in the irreversible processes: the state potential and the potential of dissipation:

➢ The state potential defines the state laws, and it can be described as the function of the state variable.

➢ The potential of dissipation is described by the associated variable which accounts for the kinetic laws. In the damage constitutive equation, the damage rate is a function of its associated variable.

In this section, only damageable materials exhibiting elastic, elastoplastic or elastoviscoplastic behaviours are considered. The damage variable is the scalar D which will be considered as a state variable amenable to the thermodynamic representation.

3.3.1 State potential

The state potential is introduced through the method of local state in the thermodynamics of irreversible process. In our study, thermal factor is taken into account, whereas it is restricted in isothermal process in Lemaitre's constitutive equations. The Helmholtz free energy is adopted and defined by the following equation, which is a convex function of the state variables.

$$\Psi = \Psi(T, \varepsilon^e, \varepsilon^p, V_k, D) \tag{3-9}$$

$$A_k = \rho \frac{\partial \Psi}{V_k} \tag{3-10}$$

where

Ψ : Helmholtz free energy

V_k : state variable

A_k : associated variable

ρ : density

In Table 3.1, state variables and their dual variables (associated variables) are listed for the behaviour of hardening and damage. The state variables can be divided into observable and internal variables. The temperature and strain tensor are observable, while others can not be measured directly and called internal variables.

Variables of thermodynamics   Table 3.1

	State Variables			Associated Variables	
		Observable	Internal		
Thermal	Temperature	T		s	Specific entropy
Elasticity	Deformation	ε		$\underline{\sigma}$	Cauchy stress tensor
Plasticity	Plastic strain		ε^p	$-\underline{\sigma}$	Cauchy stress tensor
Isotropic hardening	Accumulated plasticity		r, p	R	Isotropic strain hardening
Kinematic hardening	Back strain tensor		$\underline{\alpha}$	$\underline{X}$	Back stress tensor
Degeneration	Damage variable		D	$-Y$	Damage energy release rate

To interpret the Equ. 3-10, it can be written with the introduction of each state variable:

$$s = -\frac{\partial \Psi}{\partial T} \tag{3-11}$$

$$\underline{\sigma} = \rho \frac{\partial \Psi}{\partial \underline{\varepsilon}^e} \tag{3-12}$$

$$R = \rho \frac{\partial \Psi}{\partial r} \tag{3-13}$$

$$\underline{X} = \rho \frac{\partial \Psi}{\partial \underline{\alpha}} \tag{3-14}$$

$$Y = -\rho \frac{\partial \Psi}{\partial D} \tag{3-15}$$

When the states are coupled between mechanisms I and mechanisms J, and state variable V_I is coupled with another variable V_J. Since $A_J = \rho \frac{\partial \Psi}{V_J}$, the coupled states are:

$$\frac{\partial A_J}{\partial V_I} = \rho \frac{\partial^2 \Psi}{\partial V_I \partial V_J} \neq 0 \tag{3-16}$$

When there is no coupling among different states, it becomes:

$$\frac{\partial A_J}{\partial V_I}=\rho\frac{\partial^2\Psi}{\partial V_I\partial V_J}=0 \tag{3-17}$$

For example, there is no interaction between elastic behavior and plastic behavior:

$$\frac{\partial^2\Psi}{\partial \underline{\varepsilon}^e\partial r}=0 \tag{3-18}$$

$$\frac{\partial^2\Psi}{\partial \underline{\varepsilon}^e\partial \underline{\alpha}}=0 \tag{3-19}$$

While the damage variable is coupled with an elastic variable or variable of phase transformation, it means:

$$\frac{\partial^2\Psi}{\partial \underline{\varepsilon}^e\partial D}\neq 0 \tag{3-20}$$

Table 3.2 shows the couplings and uncoupling between different variables.

Coupling between state variables Table 3.2

	ε^p	$\underline{\alpha}$	r	T	D
ε^p	—	0	0	1	1
$\underline{\alpha}$	0	—	0	1	1
r	0	0	—	1	1
T	1	1	1	—	1
D	1	1	1	1	—

Note: 0 means uncoupling; 1 presents coupling

The free energy for each phase is divided into two parts when the elastic behavior is uncoupld with the plastic behavior:

$$\rho\Psi=\rho\Psi_e(T,\underline{\varepsilon}^e,D)+\rho\Psi_p(T,r,\underline{\alpha},D) \tag{3-21}$$

where

Ψ_e : thermoelastic free energy

Ψ_p : blocked energy by the mechanism of isotropic and kinematic hardening

When the isotropic hardening and kinematic hardening are decoupled:

$$\Psi_p=\Psi_{pi}(r,D,T)+\Psi_{pk}(\underline{\alpha},D,T) \tag{3-22}$$

Therefore, the free energy is derived as:

$$\rho\Psi=\rho\Psi_e(T,\underline{\varepsilon}^e,D)+\rho\Psi_{pi}(r,D,T)+\rho\Psi_{pk}(\underline{\alpha},D,T) \tag{3-23}$$

The thermoelastic free energy is:

$$\rho\Psi_e(T,\underline{\varepsilon}^e,D)=\frac{1}{2}(1-D)\underline{\varepsilon}^e{:}A(T){:}\underline{\varepsilon}^e+B(T){:}\underline{\varepsilon}^e+C(T)(T-T_0)^2 \tag{3-24}$$

where

$A(T)$: elastic tensor depended on Young's Module $E(T)$ and Poisson coefficient $\upsilon(T)$

$B(T)$: tensor related to the thermal expansion considered as isotropic

$\alpha(T)$: thermal dilatation coefficient

$C(T)$: scalar dependent on the purely thermal behavior

T_0 : reference temperature

The blocked energy by isotropic hardening can be expressed by:

$$\rho\Psi_{pi}=c(T)(1-D)\left[p+\frac{1}{\gamma}\exp(-\gamma p)\right] \tag{3-25}$$

The blocked energy by kinematic hardening is:

$$\rho\Psi_{pk}=\frac{1}{3}g(T)(1-D)\underline{\alpha}{:}\underline{\alpha} \tag{3-26}$$

where γ is hardening constant. $c(T)$ and $g(T)$ are hardening functions depend on material and temperature.

Therefore, the equations link between state variables and associated variables are:

$$\underline{\sigma}=\rho\frac{\partial\Psi}{\partial\varepsilon^e}=(1-D)A(T){:}\underline{\varepsilon}^e+B(T) \tag{3-27}$$

$$R=\rho\frac{\partial\Psi}{\partial p}=c(T)(1-D)[1-\exp(-\gamma p)] \tag{3-28}$$

$$\underline{X}=\rho\frac{\partial\Psi}{\partial\underline{\alpha}}=\frac{2}{3}g(T)(1-D)\underline{\alpha} \tag{3-29}$$

$$Y=-\rho\frac{\partial\Psi}{\partial D}=Y_e+Y_p \tag{3-30}$$

$$Y_e=\frac{1}{2}A(T)\underline{\varepsilon}^e \tag{3-31}$$

$$Y_p=\frac{1}{3}g(T)\alpha{:}\alpha+c(T)p \tag{3-32}$$

3.3.2 Dissipation potential

The second principle of thermodynamics or the Clausius-Duhem inequality is written as:

$$\underline{\sigma}{:}\underline{\dot{\varepsilon}}-\rho(\dot{\Psi}+s\dot{T})-\vec{q}\cdot\frac{\overrightarrow{gradT}}{T}\geqslant 0 \tag{3-33}$$

where $\vec{q}$ is the heat flux vector associated with the temperature gradient.

The rate of the Helmholtz free energy is obtained by the differentiation with respect to the state variables:

$$\dot{\Psi} = \frac{\partial \Psi}{\partial \underline{\varepsilon}^e}\dot{\underline{\varepsilon}}^e + \frac{\partial \Psi}{\partial r}\dot{r} + \frac{\partial \Psi}{\partial \underline{\alpha}}\dot{\underline{\alpha}} + \frac{\partial \Psi}{\partial D}\dot{D} \tag{3-34}$$

If the dissipation of thermodynamics is uncoupled with the dissipation of mechanics, we get:

$$\Phi = \Phi_m + \Phi_{th} \geqslant 0 \tag{3-35}$$

$$\Phi_m = \underline{\sigma}:\dot{\underline{\varepsilon}}^p - A_k\dot{V}_k + Y\dot{D} \geqslant 0 \tag{3-36}$$

$$\Phi_{th} = -\overrightarrow{grad}\cdot\vec{q}/T \geqslant 0 \tag{3-37}$$

Due to the plastic flow in the absence of damage (vice versa), the two inequalities are written as:

$$\underline{\sigma}:\dot{\varepsilon}^p - A_k\dot{V}_k = \underline{\sigma}:\dot{\underline{\varepsilon}}^p - \underline{X}:\dot{\underline{\alpha}} - R\dot{r} \geqslant 0 \tag{3-38}$$

$$-Y\dot{D} \geqslant 0 \tag{3-39}$$

Here, $-Y$ is a positive quadratic function, so that the damage $\dot{D}$ rate must be a nonnegative function.

The pseudo-potential of dissipation φ is then introduced:

$$\varphi = \varphi\left(\underline{\sigma}, A_k, \overrightarrow{grad}T, Y; \underline{\varepsilon}^e, V_k, T, D\right) \tag{3-40}$$

then:

$$\dot{V}_k = -\frac{\partial \varphi}{\partial A_k} \quad \text{and} \quad \frac{\vec{q}}{T} = -\frac{\partial \varphi}{\partial \overrightarrow{grad}T} \tag{3-41}$$

The pseudo-potential of dissipation is the sum of the thermal dissipation and viscoplasitic dissipation, which is coupled with damage:

$$\varphi = \varphi_{vp-d}\left(\underline{\sigma}, \underline{X}, R, D, T\right) + \varphi_{th}\left(\overrightarrow{grad}T, T\right) \tag{3-42}$$

$$\varphi_{th}\left(\overrightarrow{grad}T, T\right) = \frac{1}{2}k(T)\frac{\overrightarrow{grad}T}{T}\cdot\frac{\overrightarrow{grad}T}{T} \tag{3-43}$$

If the material properties of diffusion are isotropic, the Fourier law directly writes:

$$\vec{q} = -k(T)\overrightarrow{grad}T \tag{3-44}$$

The damage dissipation $\varphi_d(D, Y)$ could be chosen to be an exponential function[124]:

$$\varphi_d(D, Y) = \frac{S}{(s+1)(1-D)}\left(-\frac{Y}{S}\right)^{s+1} \tag{3-45}$$

where s and S are material coefficients characterizing the ductile damage evolution: they are functions of temperature T.

In order to interpret the viscoplastic-damage dissipation, a yield function (plasticity convex) f and a plastic potential (non associative theory) F can be introduced[168], for instance, as following:

$$f = \frac{J_2\left(\underline{\sigma} - \underline{X}\right)}{\sqrt{1-D}} - \frac{R}{\sqrt{1-D}} - \sigma_y(T) \tag{3-46}$$

$$F = f + \frac{a(T)}{2(1-D)} J_2^2\left(\underline{X}\right) + \frac{b(T)}{2(1-D)} R^2 + \frac{S(T)}{[s(T)+1](1-D)} \left[-\frac{Y}{S(T)} \right]^{s(T)+1} \tag{3-47}$$

where $a(T)$ and $b(T)$ are material coefficients dependent on the temperature to describe the nonlinear isotropic and kinematic hardening respectively.

The scalar J_2 is defined as following:

$$J_2\left(\underline{\sigma}\right) = \sqrt{\frac{3}{2} \underline{\sigma}^D : \underline{\sigma}^D} \tag{3-48}$$

$$J_2\left(\underline{\sigma} - \underline{X}\right) = \sqrt{\frac{3}{2} \left(\underline{\sigma}^D - \underline{X}^D\right) : \left(\underline{\sigma}^D - \underline{X}^D\right)} \tag{3-49}$$

$$\underline{\sigma}^D = \underline{\sigma} - \frac{1}{3} Tr\left(\underline{\sigma}\right) \cdot I \tag{3-50}$$

The accumulated plastic strain p is defined by its rate:

$$\dot{p} = \sqrt{\frac{2}{3} \underline{\dot{\varepsilon}}^p : \underline{\dot{\varepsilon}}^p} \tag{3-51}$$

Therefore, the viscoplastic-damage dissipation is written as:

$$\varphi_{vp-d} = \varphi_{vp-d}\left(\underline{\sigma}, \underline{X}, R, Y, D, T\right) = \frac{\mu(T)}{1 + n(T)} < F >^{n(T)+1} \tag{3-52}$$

where $\mu(T)$ and $n(T)$ are coefficients of viscosity. $<\ >$ presents the Heaviside step function, also called unit step function, which is a discontinuous function whose value is zero for negative argument and one for positive argument.

For specific materials (the generalized normality), we can introduce the Lagrange multiplier $\dot{\lambda}$ which leads to the flux of state variables:

$$\underline{\dot{\varepsilon}}^p = \dot{\lambda} \frac{\partial f}{\partial \underline{\sigma}} = \frac{3}{2} \frac{\dot{\lambda}}{\sqrt{1-D}} \frac{\underline{\sigma}^D - \underline{X}^D}{J_2\left(\underline{\sigma} - \underline{X}\right)} \tag{3-53}$$

$$\underline{\dot{\alpha}} = -\dot{\lambda} \frac{\partial F}{\partial \underline{X}} = \frac{3}{2} \frac{\dot{\lambda}}{\sqrt{1-D}} \frac{\underline{\sigma}^D - \underline{X}^D}{J_2\left(\underline{\sigma} - \underline{X}\right)} - a \dot{\lambda} \underline{\alpha} \tag{3-54}$$

$$\dot{r} = -\dot{\lambda}\frac{\partial F}{\partial R} = \dot{\lambda}\left(\frac{1}{\sqrt{1-D}} - br\right) \tag{3-55}$$

$$\dot{D} = -\dot{\lambda}\frac{\partial F}{\partial Y} = \frac{\dot{\lambda}}{1-D}\left(-\frac{Y}{S}\right)^{s} \tag{3-56}$$

For viscoplasticity, the Lagrange multiplier is[168]:

$$\dot{\lambda} = \dot{\lambda}_{vp} = \left\langle\frac{f}{k}\right\rangle^{n} \tag{3-57}$$

where k and n characterize the material viscosity.

3.3.3 Triaxiality and damage equivalent stress

To formulate an isotropic criterion, it is postulated that the mechanism of damage is governed by the total elastic strain energy, which consists of distortion energy and volumetric energy[124]:

$$\omega^{e} = \omega_{D}^{e} + \omega_{H}^{e} \tag{3-58}$$

$$\omega^{e} = \int_{0}^{\varepsilon^{e}} \underline{\sigma}: \mathrm{d}\underline{\varepsilon}^{e} = \int_{0}^{\varepsilon_{D}^{e}} \underline{\sigma}^{D}: \mathrm{d}\underline{\varepsilon}_{D}^{e} + 3\int_{0}^{\varepsilon_{H}^{e}} \sigma_{H}: \mathrm{d}\varepsilon_{H} \tag{3-59}$$

Hooke's law coupled with damage contains the deviatoric and the hydrostatic parts:

$$\varepsilon_{D}^{e} = \frac{1+\upsilon}{E}\frac{\underline{\sigma}^{D}}{1-D}, \varepsilon_{H} = \frac{1-2\upsilon}{E}\frac{\sigma_{H}}{1-D} \tag{3-60}$$

Therefore, Equ. 3-59 changes to:

$$\omega^{e} = \frac{1}{2}\left(\frac{1+\upsilon}{E}\frac{\underline{\sigma}^{D}:\underline{\sigma}^{D}}{1-D} + 3\frac{1-2\upsilon}{E}\frac{\sigma_{H}^{2}}{1-D}\right) \tag{3-61}$$

the von Mises equivalent stress is defined as:

$$\sigma_{eq} = \left(\frac{3}{2}\underline{\sigma}^{D}:\underline{\sigma}^{D}\right)^{1/2} \tag{3-62}$$

Similar to the equivalent stress in plasticity, the damage equivalent stress is defined by that this energy in a multiaxial state is equal to that in an equivalent uniaxial state[124].

$$-Y = \frac{\omega^{e}}{1-D} = \frac{\sigma_{eq}^{2}}{2E(1-D)^{2}}\left[\frac{2}{3}(1+\upsilon) + 3(1-2\upsilon)\left(\frac{\sigma_{H}}{\sigma_{eq}}\right)^{2}\right] \tag{3-63}$$

$$\sigma^{*} = \sigma_{eq}\left[\frac{2}{3}(1+\upsilon) + 3(1-2\upsilon)\left(\frac{\sigma_{H}}{\sigma_{eq}}\right)^{2}\right]^{1/2} \tag{3-64}$$

where

$\frac{\sigma_{H}}{\sigma_{eq}}$: triaxiality ratio, which plays an important role in damage evolution.

The term in square brackets is called triaxiality function $f\left(\frac{\sigma_H}{\sigma_{eq}}\right)$:

$$f\left(\frac{\sigma_H}{\sigma_{eq}}\right)=\frac{2}{3}(1+\upsilon)+3(1-2\upsilon)\left(\frac{\sigma_H}{\sigma_{eq}}\right)^2 \tag{3-65}$$

When a triaxiality ratio $\sigma_H/\sigma_{eq}=1/3$, it leads to $f\left(\frac{\sigma_H}{\sigma_{eq}}\right)=1$.

From the above formulations, the damage equivalent stress differs from the von Mises equivalent stress by the triaxiality function. Plasticity is connected to slips, which are caused by shear stress. There is no dependency on hydrostatic observed. Damage is connected with debonding which is influenced by hydrostatic or triaxiality ratio. In our following experiment of notched specimen, the radius of notch controls the distributions of triaxiality ratio, thus identifies damage parameters by using various riaxiality ratios.

3.3.4 Threshold and critical damage

From a microscopic point of view, ductile plastic damage consists of the nucleation, growth and coalescence of cavities induced by larger plastic strains. Section 3.2.4 presented that the variable D could be measured by variations of elasticity modulus. From the experiment of many metallic materials subjected to a uniaxial monotonically increasing load, it is noticed that no damage takes place when strain is smaller than some specific value; when the accumulated strain reaches a large enough value, rupture initiates. Such phenomena can be summarized in a schematic diagram of damage evolution (Figure 3.3), and three main damage parameters are threshold damage strain ε_D, failure strain ε_R, and critical damage D_c.

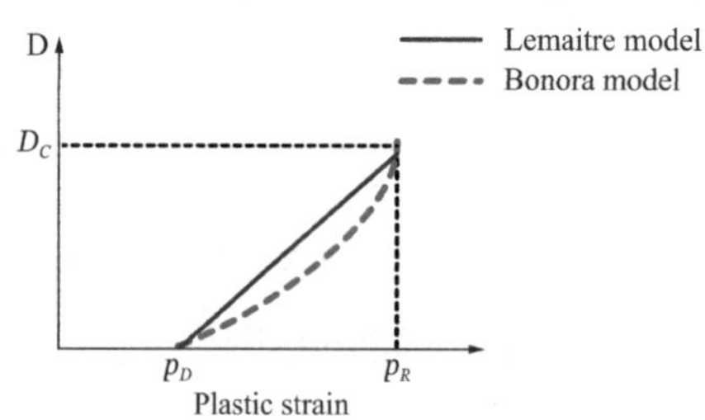

Figure 3.3 Schematic diagram of damage evolution and damage parameters

Starting from the dissipation potential, we can deduce the threshold and critical damage of a multiaxial model from experimental results of uniaxial model in the framework of thermodynamics. The Equ.3-64 can be written as:

$$-Y=\frac{K^2}{2E}f\left(\frac{\sigma_H}{\sigma_{eq}}\right) \tag{3-66}$$

with $\frac{\sigma_{eq}}{1-D}=K=\text{const.}$

With the hypotheses of isotropic damage and isotropic hardening, in Lemaitre-Chaboche model, damage ratio is written as:

$$\dot{D} = -\dot{\lambda}\frac{\partial F}{\partial Y} = \left(-\frac{Y}{S}\right)^{s}\dot{p} \tag{3-67}$$

Inserting Equ.3-66 to the above equation, we obtain:

$$\dot{D} = \left(\frac{K^2}{2ES}\right)^{s} f\left(\frac{\sigma_H}{\sigma_{eq}}\right)^{s}\dot{p} \tag{3-68}$$

In integral form, it is written as:

$$D = \left(\frac{K^2}{2ES}\right)^{s} f\left(\frac{\sigma_H}{\sigma_{eq}}\right)^{s}\langle p - p_D\rangle \tag{3-69}$$

The simplification of fracture condition is: $p = p_R$ and $D = D_c$.

Therefore, the Equ.3-69 is written as:

$$p_R - p_D = \left(\frac{K^2}{2ES}\right)^{-s} f\left(\frac{\sigma_H}{\sigma_{eq}}\right)^{-s} D_c \tag{3-70}$$

In uniaxial condition, $\sigma_H/\sigma_{eq} = 1/3$ and it is assumed that no distinction exists between total strain ε and plastic strain p and $p_R/p_D = \varepsilon_R/\varepsilon_D$.

$$p_R\left(\frac{1}{3}\right) = \varepsilon_R = \left(\frac{K^2}{2ES}\right)^{-s}\frac{D_c}{1-\varepsilon_D/\varepsilon_R} \tag{3-71}$$

In multiaxial case, we get:

$$p_R\left(\frac{\sigma_H}{\sigma_{eq}}\right) = f\left(\frac{\sigma_H}{\sigma_{eq}}\right)^{-s}\varepsilon_R = \left(\frac{K^2}{2ES}\right)^{-s} f\left(\frac{\sigma_H}{\sigma_{eq}}\right)^{-s}\frac{D_c}{1-\varepsilon_D/\varepsilon_R} \tag{3-72}$$

3.4 Ductile damage models

3.4.1 Introduction

The damage is called ductile when it occurs simultaneously with plastic strains. It results from the nucleation of cavities due to decohesions between inclusions and the matrix followed by their growth and their coalescence through the phenomenon of plastic instability. In plastic deformations induced by the welding process, the principal mechanism of failure is due to the ductile damage.

Considerable studies have been carried out in the topic of ductile damage, and many different models have been proposed. All these formulations in the literature could be categorized in three main categories:

➢ The first approach is the abrupt failure criteria. Failure is predicted to occur when one external variable, which is uncoupled from other internal variables, reaches its critical value. McClintock[138] firstly recognized the role of microvoids in ductile failure process and tried to correlate the mean radius of the nucleated cavities to the overall plastic strain increment. Rice and Tracey[160] analytically studied the evolution of spherical voids in an elastic-perfectly plastic matrix. In these pioneering studies, the interaction between microvoids, the coalescence process and hardening effects were neglected and failure was postulated to occur when the cavity radius reachs a critical value specific for the material.

➢ Second, porous solid plasticity. The effect of ductile damage, as proposed by Gurson[87] and Rousselier[163], was taken into account in the yield condition by a porosity term that progressively shrinks the yield surface. Later, Needleman and Tvergaard[142] as well as Koplik and Needleman[110] extended the initial formulation proposed by Gurson in order to include the acceleration in the failure process induced by void coalescence (GTN model). More recently, a number of finite element unit cell based micromechanical studies have been performed in order to correlate voids evolution and interaction with the resulting macroscale material yield function. Tvergaard and Niordoson[190] investigated the role of smaller sized voids in a ductile damage material by using a nonlocal plasticity model as proposed by Acharya and Bassani[5]. Schacht et al.[170], used the 3D voided unit cell based approach to investigate the role and the effects associated with the crystallographic orientation of the matrix material, and they found a substantial dependency of the growth and coalescence phase with the anisotropy of the material surrounding the voids.

➢ The third is continuum damage mechanics (CDM). In this approach, damage is assumed to be one of the internal variables that accounts for the effects on the material constitutive response induced by the irreversible processes that occur in the material microstructure. As the starting point of continuum damage mechanics (CDM), Kachanov was the first to introduce a continuous damage variable in 1958[103]. Then, the concept of effective stress was introduced by Rabotnov in 1968[158]. The CDM framework for ductile damage was later developed by Lemaitre and Chaboche[124][125].

More recently, starting from the consideration that the gradient effect is important when the characteristic dimension of the plastic strain or damage is of the same order of the material intrinsic length scale, a number of so-called non-local theories have been proposed by Fleck et al.[72][73] and Bammann et al.[13].

Relative to the initial framework proposed by Lemaitre, several damage models, based on the use of special expressions for the damage dissipation potential, have been derived by different authors[31][180][181]. In addition, Bonora[18][19][20] proposed a damage model formulation in the framework of CDM. This formulation is material independent, and it allows the description of different damage evolution laws with plastic strain without the need to change the choice of the damage dissipation potential. Later, Pirondi and Bonora[154] reformulated this model, and pointed out how the experimentally determined flow stress curve does not require the substitution of the effective stress in its expression.

In terms of the first approach, it is so simple that it is far away from the reality in many situations, thus leading to the limited application. In the following section, we will introduce the later two branches of approaches, which are developed based on the framework of irreversible thermodynamics.

3.4.2 Models based on porous solid plasticity

Gurson[87][188][39]

Concerned with the ductile failure of porous materials, Gurson's model[87] is widely accepted. Assuming that voids are initially spherical and remain spherical in the growth process, the approach to describe the micromechanical effects of damage in ductile metals was first proposed by Gurson and then it was extended to GTN model by Tvergaard and Needleman[188]. The model is interpreted as an extension of conventional von Mises plasticity, and it taks into account the experimental observation that ductile degradation processes consist of the nucleation, the growth and the coalescence of microvoids.

The plastic flow with cavities proposed by Gurson rises from a plastic potential of form[87]:

$$\Phi = \left(\frac{\sigma_{eq}}{\sigma_0}\right)^2 + 2f\cosh\left(\frac{3}{2}\frac{\sigma_H}{\sigma_0}\right) - (1 + f^2) \tag{3-73}$$

with

$$\sigma_H = \frac{1}{3}Tr(\underline{\sigma}) = \frac{1}{3}J_1(\underline{\sigma})$$

$$\sigma_{eq} = \sqrt{\frac{3}{2}\underline{S}:\underline{S}} = J_2(\underline{\sigma})$$

The above flow potential Φ characterizes the porosity in terms of a single scalar internal variable, where f is the void volume fraction defined by:

$$f = \frac{V_t - V_0}{V_t} = 1 - \frac{\rho}{\rho_0} = \frac{J-1}{J} \tag{3-74}$$

$$J = \frac{\rho_0}{\rho} = \frac{V_t}{V_0} \tag{3-75}$$

where

V_0 : volume of original matrix

V_t : volume of matrix at time t

ρ_0 : density of original matrix

ρ : density of matrix at time t

It is noticeable that when there is no cavity in matrix, $V_t = V_0 \rightarrow f = 0$ and if the matrix is totally ruptured, $\rho \rightarrow 0$ and $f \rightarrow 1$.

In GTN model, the damage variable f is replaced by f^* and three additional parameters (q_1, q_2, q_3) are introduced[188]. The yield function is written as:

$$\Phi = \left(\frac{\sigma_{eq}}{\sigma_{YM}}\right)^2 + 2q_1 f^* \cosh\left(\frac{3}{2}\frac{q_2\sigma_H}{\sigma_{YM}}\right) - (1 + q_3 f^{*2}) \tag{3-76}$$

$$f^* = \begin{cases} f & \forall f \leqslant f_c \\ f + \dfrac{q_1^{-1} - f_c}{f_F - f_c}(f - f_c) & \forall f > f_c \end{cases} \tag{3-77}$$

where the parameter f_c characterizes the beginning of void nucleation and f_F denotes the final failure. The parameters q_1, q_2 and q_3 make the predictions of the Gurson model agree with his numerical studies of materials containing periodically distributed circular cylindrical and spherical voids.

The increasing rate of the microvoid volume fraction is given by:

$$\dot{f} = \dot{f}_n + \dot{f}_g \tag{3-78}$$

where $\dot{f}_n$ and $\dot{f}_g$ are, respectively, the nucleation and the growth rate of microvoids.

Assuming that the matrix material is plastically incompressible, the rate of increasing of the microvoid volume fraction, due to the growth of existing microvoids, is given by:

$$\dot{f}_g = (1 - f)Tr(\dot{\varepsilon}^p) \tag{3-79}$$

The nucleation of voids is a very complex physical process depending on the microstructure of the material. Chu and Needleman[39] improved the Gurson model by a statistical approach:

$$\dot{f}_n = \frac{f_N}{s_N\sqrt{2\pi}} \times \exp\left[-\frac{1}{2}\left(\frac{\varepsilon_{eq}^p - \varepsilon_N}{s_N}\right)^2\right]\dot{\varepsilon}_{eq}^p \tag{3-80}$$

The hardening of the matrix material depends on the equivalent plastic strain ε_{eq}^p and it is given by a power law including three material parameters:

$$\sigma_{YM}^p(\varepsilon_{eq}^p) = \sigma_0\left(\frac{\varepsilon_{eq}^p}{\varepsilon_0} + 1\right)^n \tag{3-81}$$

The initial yield stress σ_0 and ε_0 and n are hardening parameters. Using the equivalence of microscopic and macroscopic plastic work, one obtains an evolution equation for the internal hardening variable:

$$\dot{\varepsilon}_{eq}^p = \dot{\lambda}\frac{1}{\sigma_{YM}(1-f)}\underline{\sigma}:\frac{\partial \Phi}{\partial \underline{\sigma}} \tag{3-82}$$

where $\dot{\lambda}$ is the plastic multiplier.

In the represented form, the GTN model contains 12 material parameters:

$$P = (\sigma_0,\ \varepsilon_0,\ n,\ q_1,\ q_2,\ q_3,\ f_0,\ f_c,\ f_F,\ f_N,\ \varepsilon_N,\ s_N) \tag{3-83}$$

Gurson model was studied and then extended by many researchers. After Tvergaard[188] uses empirical parameters to take into account cavity growth and the cavities coalescence, the Richmond and Smelser[161] criterion introduces a parameter that considers shear bands, which could dominate the plastification process. The Sun and Wang[178] criterion is the only one to be based on an inner approach of the Gurson model under a macroscopic stress on the hollow sphere boundary. It is noticed that many processed materials, such as rolled plates, have non-spherical voids. Even for materials having initially spherical voids, the voids will change to prolate or oblate shape after deformation, depending on the state of the applied stress. Because of the limited application of Gurson model due to the assumption of spherical voids, Gologanu et al.[81][82][83] extended the Gurson model to the spheroidal void incorporating void shape effects, namely the GLD (Gologanu-Leblond-Devaux) model. The GLD model provides an important improvement to the widely adopted GT model in describing void growth and the corresponding material behavior during the ductile fracture process[149][76]. Brunet[24][25]extended GLD model for prolate voids, plane-stress state and non-linear kinematic hardening including initial quadratic anisotropy of the matrix or base material. In order to predict the damage evolution induced by porosity increasing in bulk forming processes, Staub and Boyer[173] proposed a void growth model, based upon the Rice and Tracey analysis and suited for thermo-elasto-plastic finite-element modelling.

Rousselier[163][164][165]

Although Rousselier model[163][164] was proposed in the form of a thermodynamically based continuum damage mechanics method, we classify it into the section of porous solid damage (section 3.4.2), because this ductile fracture model is related to void nucleation, growth, and coalescence in metals. The Rousselier criterion[165] is based on a macroscopic thermodynamic approach of irreversible phenomena under the local equilibrium assumption, i.e. the state of a system can be defined locally using the same variables as at the equilibrium state. The parameter values of this criterion must be determined experimentally.

In the Rousselier model, several hypotheses are adopted:

➢ Hypothesis 1: The internal hardening p and damage variable β are scalar, and the associated variables with p and β are P and B.

➢ Hypothesis 2: In the free energy, the variables are separated. The thermodynamic potential Φ is of the from:

$$\Phi = \Phi_e(\varepsilon^e) + \Phi_p(p) + \Phi_\beta(\beta) \tag{3-84}$$

➢ Hypothesis 3: The plastic potential F is of the form:

$$F = F_1(\tilde{\sigma}_{eq}, P) + F_2(\tilde{\sigma}_m, B) \tag{3-85}$$

with

$$F_1 = \tilde{\sigma}_{eq} + P(p)(\text{von Mises yield criterion}) \tag{3-86}$$

$$F_2 = B(\beta)g(\tilde{\sigma}_m) \tag{3-87}$$

$$g(\tilde{\sigma}_m) = K\exp\left(\frac{\tilde{\sigma}_m}{\sigma_1}\right) \tag{3-88}$$

in which K is the constant of integration, and σ_1 is a constant too.

Based on the concepts of "principle of simplicity" and "standard model", finally, the plastic potential is written as:

$$\Phi = \tilde{\sigma}_{eq} - R(p) + B(\beta)K\exp\left(\frac{\tilde{\sigma}_m}{\sigma_1}\right) \tag{3-89}$$

where $R(p)$ is the hardening curve of the material; p is the hardening variable (the cumulated plastic strain); β is the damage variable. Both p and β are scalar.

The damage variable β is defined by:

$$\dot{\beta} = \frac{\dot{\upsilon}}{\upsilon} \tag{3-90}$$

where υ is the average volume of the cavities.

The function $B(\beta)$ is defined by:

$$B(\beta) = \frac{\sigma_1 f_0 \exp(\beta)}{1 - f_0 + f_0 \exp(\beta)} = \sigma_1 f \tag{3-91}$$

where f_0 is a constant including the initial volume fraction of cavities, and f is the actual volume fraction of cavities.

In Rousselier model, it did not introduce a critical value of damage variable because the stresses decrease abruptly and vanish, and zone of strain and damage localization were assimilated to a crack. Therefore, three parameters are introduced:

f_0, related to the initial volume of inclusions.

σ_1, related to the resistance of the metal matrix to the growth and coalescence of cavities.

D, does not depend on the material and can be taken to be equal to 2 , at least for the initial volume fraction of f_0 equal to or smaller to than 10^{-3}.

The Rousselier model can be extended to account for void nucleation effects and by introducing a void volume fraction f_F at which failure occurs[165].

In addition, based on the study on voids and cracks of the porous materials, numerous models have been developed, such as Kim and Kwon model[106]. More recently, Bigoni and Piccolroaz[17] have proposed a macroscopic yield function to agree with a variety of experimental data relative to soil, concrete, rock, metallic and composite powders, metallic foams, porous materials and polymers. Their yield function was presented as a generalization of several classical criteria, and it also gives an approximation of the Gurson criterion. Trillat and Pastor[186][187] analyzed Gurson's model, and proposed a yield criterion for porous media with spherical voids to satisfy Gurson criterion conditions. Enakoutsa and Leblond[64] made some improvements in the numerical aspect and algorithm for the assessment of a phenomenological nonlocal variant of Gurson's model suggested by Leblond et al.[120].

3.4.3 Models based on continuum damage mechanics

Lemaitre & Chaboche[122][123][124][125][29]

There is no doubt that continuum damage mechanics (CDM) is connected to the names of

Lemaitre and Chaboche. Section 3.3 has given a detailed description of the thermodynamics of isotropic damage founded by Lemaitre and Chaboche, from the development of potential to the definition of threshold and critical damage in the framework of irreversible thermodynamics. The formulation of Lemaitre damage model can be found there.

The multiaxial ductile plastic damage strain model is as following:

In differential form:

$$\dot{D} = \frac{D_c}{p_R - p_D}\left[\frac{2}{3}(1+\upsilon) + 3(1-2\upsilon)\left(\frac{\sigma_H}{\sigma_{eq}}\right)^2\right]\dot{p} \tag{3-92}$$

In integrated form:

$$D \approx \frac{D_c}{p_R - p_D}\left\{p\left[\frac{2}{3}(1+\upsilon) + 3(1-2\upsilon)\left(\frac{\sigma_H}{\sigma_{eq}}\right)^2\right] - p_D\right\} \tag{3-93}$$

In the represented form, besides material parameter υ (Poisson's ratio), Lemaitre model contains 3 material damage parameters (see as Figure 3.3):

$$P = (p_R,\quad p_D,\quad D_c) \tag{3-94}$$

Identification of this model consists of the quantitative evaluation of the three coefficient characteristics by mean of tensile tests.

Further, Lemaitre and Chaboche[124] introduced the damage variable into the viscoplastic model. The rate of creep damage can be written as:

$$D = \left\langle\frac{\chi(\sigma)}{A}\right\rangle^r (1-D)^{-k\langle\chi(\sigma)\rangle} \tag{3-95}$$

where, $\chi(\sigma)$ is equivalent stress on creep fracture. A, r and k are three characteristic creep damage coefficients for the material.

Bonora[18][19][20]

Bonora[18][19] proposed a damage model in the framework of CDM. This formulation is material independent, allowing the description of different damage evolution laws with plastic strain without the need to change the choice of the damage dissipation potential.

In Bonora's work[18][20], the following expression for the damage dissipation potential was proposed:

$$F_D = \left[\frac{1}{2}\left(-\frac{Y}{S_0}\right)^2 \frac{S_0}{1-D}\right]\frac{(D_c - D)^{(\alpha-1)/\alpha}}{p^{(2+n)/n}} \tag{3-96}$$

The associated damage variable Y is expressed by:

$$Y=-\frac{\sigma_{eq}^2}{2E(1-D)^2}f\left(\frac{\sigma_H}{\sigma_{eq}}\right) \tag{3-97}$$

The triaxiality function $f(\sigma_H/\sigma_{eq})$ is deified by Equ. 3-65.

If the von Mises stress of a duticle material can be guided by Ramberg-Osgood power law[159]as:

$$\frac{\sigma_{eq}}{(1-D)}=Kp^{1/n} \tag{3-98}$$

where K is a material constant, we have:

$$\frac{\partial F_D}{\partial Y}=-\left(\frac{\sigma_{eq}^2}{(1-D)^2}\right)f\left(\frac{\sigma_H}{\sigma_{eq}}\right)\frac{1}{2ES_0}\frac{(D_c-D)^{(\alpha-1)/\alpha}}{p^{(2+n)/n}}\frac{1}{1-D} \tag{3-99}$$

Finially, damage rate is given in differential form:

$$\dot{D}=-\dot{\lambda}\frac{\partial F_D}{\partial Y}=\alpha\frac{(D_c-D_0)^{1/\alpha}}{\ln(p_R/p_D)}\left[\frac{2}{3}(1+\upsilon)+3(1-2\upsilon)\left(\frac{\sigma_H}{\sigma_{eq}}\right)^2\right](D_c-D)^{(\alpha-1)/\alpha}\frac{\dot{p}}{p} \tag{3-100}$$

Given the assumption of proportional loading, the damage evolution is derived as:

$$D=D_0+(D_c-D_0)\left\{1-\left[1-\frac{\ln(p/p_D)}{\ln(p_R/p_D)}f\left(\frac{\sigma_H}{\sigma_{eq}}\right)\right]^{\alpha}\right\} \tag{3-101}$$

where α is the damage exponent, D_0 and D_c present respectively the initial and critical damage. The rupture strain is p_R. The p_D is the threshold strain. Here no distinction is made between total strains and plastic strains.

Compared with Lemaitre model, the constitutive equation of damage proposed by Bonora contains two additional parameters: D_0 and α. It includes 5 material parameters:

$$P_i=(\varepsilon_{th},\quad \varepsilon_f,\quad D_0,\quad D_c,\quad \alpha) \tag{3-102}$$

Saanouni[35][167][168]

Saanouni proposed a fully coupled isotropic and isothermal formulation by using a single yield surface both for plasticity and damage. The dual (or force) variables $(\underline{\sigma},\underline{X},R,Y)$ are derived from the state potential defined as the Helmholtz free energy while the flux variables $(D,\underline{\dot{\alpha}},\dot{r},\dot{D})$ are derived from an appropriated dissipation potential $F(\underline{\sigma},\underline{X},R,Y,D)$ together with a single yield function $f(\underline{\sigma},\underline{X},R,D)$ in the framework of non-associative theory. The complete sets of state constitutive equations are given as following:

Helmholtz free energy

$$\rho\Psi=\frac{1}{2}(1-D)\underline{\varepsilon}^e:(2\mu\underline{1}+\lambda_e\underline{1}\otimes\underline{1}):\underline{\varepsilon}^e+\frac{1}{3}C(1-D)\alpha:\alpha+\frac{1}{2}Q(1-D)r^2 \tag{3-103}$$

Plastic potential

$$F = f + \frac{3}{4}\frac{a}{C(1-D)}X{:}X + \frac{1}{2}\frac{b}{Q(1-D)}R^2 + \frac{1}{(1-D)^\beta}\frac{A}{\gamma+1}\left(\frac{Y}{A}\right)^{\gamma+1} \tag{3-104}$$

Yield function in the stress space

$$f(\underline{\sigma}, \underline{X}, R) = \frac{\left\|\underline{\sigma} - \underline{X}\right\|_s - R}{\sqrt{1-D}} - \sigma_y < 0 \tag{3-105}$$

Kinetic law of damage evolution

$$\dot{D} = -\dot{\lambda}\frac{\partial F}{\partial Y} = \frac{\dot{\lambda}}{(1-D)^\beta}\left(-\frac{Y}{A}\right)^\gamma \tag{3-106}$$

where ρ is the material density in the current undamaged configuration. λ and μ are the classical Lame's constants. Q the scalar isotropic hardening modulus, C the kinematic hardening modulus, a the kinematic hardening non-linearity coefficient, b the isotropic hardening non-linearity coefficient, γ, A and β are the material coefficients characterizing the ductile damage evolution.

Saanouni's model is widely applied in the metal process of forming.[168] gives an exhaustive presentation of the theoretical, numerical and geometrical aspects of the virtual metal forming with ductile damage.

After the foundation and development of damage model, another important work is the study on assessment and comparison of different kinds of models. For example, Drabek and Bohm compared Gurson model and Rousselier model[60][61]. Hambli[91] did comparative study between Lemaitre and Gurson damage models in crack growth simulation. Lammer and Tsamakis[113] compared different CDM models with reference to homogeneous and inhomogeneous deformations providing a formulation generalization to finite deformation.

3.4.4 Anisotropic damage

Second order damage tensors have been introduced e.g. by Murakami and Ohno[140], Betten[16], Krajcinovic[111], Chow and Wang[38], Murakami[141], Lu and Chow[131], Bruhns and Schiesse[22], and Steinmann and Carol[174]. Brunig[26] proposed an anisotropic CDM model by using the porosity as a definition for the damage variable. The fourth rank damage tensors were already introduced, for example, in the anisotropic creep damage modelling by

Chaboche[30].

The simplest way to model anisotropic damage is to assume that damage will only occur in the plane perpendicular to the highest principal stress[121]:

$$[\tilde{\sigma}] = \begin{bmatrix} \sigma_1 & 0 & 0 \\ 0 & \sigma_2 & 0 \\ 0 & 0 & \dfrac{\sigma_3}{1-D} \end{bmatrix} \tag{3-107}$$

where $\sigma_1 < \sigma_2 < \sigma_3$

Murakami[141] has developed an approach based on geometrical considerations with a second order damage tensor. In his model, not only the effective cross section is decreasing during loading, but also the orientation of that cross section. Assuming that the shape of the considered cross section is not changing (this leads to the case of orthotropic damage) we can write according to the isotropic case:

$$(\mathbf{1} - \mathbf{D})\vec{n}\delta S = \tilde{n}\delta\tilde{S} \tag{3-108}$$

where $\mathbf{1}$ is the second order identity tensor, and $\mathbf{D}$ is the second order damage tensor.

When the principal directions of stress and damage coincide, the following expression is valid:

$$[\tilde{\sigma}] = \begin{bmatrix} \dfrac{\sigma_1}{1-D_1} & 0 & 0 \\ 0 & \dfrac{\sigma_2}{1-D_2} & 0 \\ 0 & 0 & \dfrac{\sigma_3}{1-D_3} \end{bmatrix} \tag{3-109}$$

3.4.5 Conclusion

In terms of metallic metrials, it is not necessary to use the complicated anisotropic damage under the condition of RVE size in mesoscopic approach. Thus, the isotropic damage is chosen to develop the damage model.

Viscoelasticity is preponderant in polymers and all their compounds (especially organic matrix composites), it is of little importance for metals and alloys. Viscoelasticity is less important for welding due to the short time of the welding process. Therefore, the coupling between damage and elastoplasticity can describe the damage behaviour induced by welding.

3.5 Welding damage

3.5.1 Introduction

Macro cracks are observed in heat affected zone (HAZ) or welding line of metallic materials under specific welding condition. In addition, because nonmetallic inclusions in a metal can be the sites of void nucleation, and welding process deepens on such segregation of non metallic elements, and causes void nucleation, growth and coalesence, at the end leads to ductile fracture of the structure[197]. There are also many small or micro defects in the welding component induced by many factors such as material ingredient fluctuates, generation of hydrogen, inclusion of impurity, etc.

3.5.2 Heat affected zone (HAZ) of weld

Usually, a weldment is made of three main areas (Figure 3.4):

➢ The weld metal deposed during the welding operation.

➢ The heat affected zone (HAZ) where the microstructure of the base metal is affected by local heating due to the deposit of the weld metal in the liquid state (i.e. at its melting temperature).

➢ The base metal.

In addition, the HAZ of weldments consists of various microstructural states (mainly three) in a narrow zone. In fact, the welding procedure induces local heating of the base metal at temperatures ranging from 1400°C at the fusion line to 750°C in the so called heat affected zone (HAZ). Micro structural heterogeneity in the HAZ can be related to phase transformations. The schematic diagram of phase transformation is given in Figure 2.4. There the initial and end temperatures of austenite transformation are respectively denoted as A_{c_1} and A_{c_3}. The three main parts of the HAZ are the following (Figure 3.4):

➢ Coarse grained heat affected zone (CGHAZ): $T_{\max} \gg A_{c_3}$, the austenite transformation is complete and austenite grain growth is promoted by elevated temperature (Nearest zone from the fusion line).

➢ Fine grained heat affected zone (FGHAZ): T_{max} is just above A_{c_3}, the austenite transformation is nearly complete but austenite grain growth is limited.

➢ Inter critical heat affected zone (ICHAZ): $A_{c_1} < T_{max} < A_{c_3}$, the martensite is partially transformed into austenite during the welding thermal cycle.

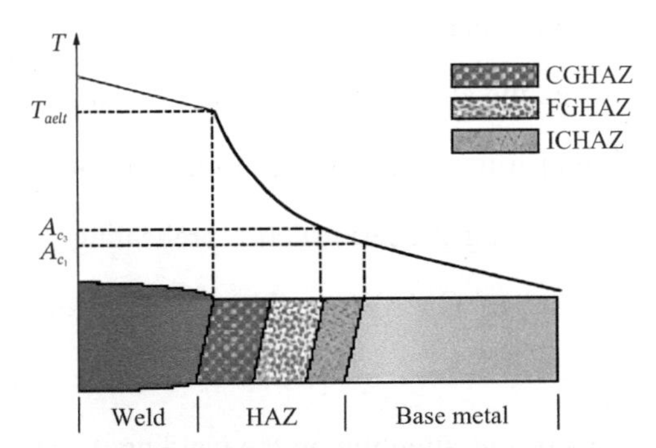

Figure 3.4 Schematic diagram of different zones of a weld joint

3.5.3 Damage and cracking induced by welding

Basing on the study on sensitivity of weldments to cracks, there are different types of cracking that may appear due to the welding procedure:

➢ Solidification cracking: occurs for metals which have a high thermal expansion coefficient.

➢ Hot cracking: occurs where the solidus temperature is reached as a liquid film may be trapped between the solidification limit and the columnar grain. When films are unable to accommodate the tensile contraction strains, cracking occurs. Generally, the austenite former elements are said to increase the hot cracking susceptibility and ferrite former elements are said to lower it. Barnes et al.[14] studied the effect of Nb, Mn and welding conditions on hot cracking. They found that for high Nb content, local segregation may lead to high local concentration of Nb , and a liquid film which is an Nb intermetallic phase very sensitive to cracking can be formed (if no crack occurs, this intermetallic phase is solidified in δ ferrite which is retained during cooling). On the other hand, Mn increases the resistance to hot cracking. From, their study, Barnes et al. also[14] concluded that the welding parameters have little influence on the hot cracking susceptibility.

➢ Reheat cracking: may occur for two reasons. The principal cause is that when heat treating susceptible steels, the grain interior becomes strengthened by carbide precipitation, forcing the relaxation of residual stresses by creep deformation at the grain boundaries. The second factor of reheat cracking is the segregation of elements at grain boundaries. It is also called reverse temper embrittlement. To investigate the weldment sensibility to reheat cracking due to temper embrittlement, a step cooling, which is a multi-stages tempering treatment, is often performed.

➢ Hydrogen cracking: is possible if hydrogen is dissolved in the weld metal during the welding procedure. Indeed, when the austenite martensite transformation proceeds, the hydrogen solubility strongly decreases and it may leads to the formation of hydrogen bubbles. However, applying some procedures of welding process, such a preheating process, a post welding heat treatment, a TIG welding instead of GMAW, could minimized the risk of hydrogen induced cracking (HIC).

In our study, our model focus our attention on the welding damage and cracking induced by welding restraint, stress of phase transformation and viscoelastoplastity. The factors of Hydrogen, oxidation and corrosion are neglected.

3.6 Summary

In this chapter, definitions of damage variable and thermodynamics of isotropic damage are presented in the framework of irreversible thermodynamics. Various widely used models are studied, and these models are foundation of the coupled-transformation damage constitutive equations.

Our study herein is based on the isotropic damage and CDM. Although anisotropy is closer to reality of most materials, the assumption of isotropic damage is realistic in many cases for metallic materials, especially under proportional loading when the directions of principal stress remain constant.

In Chapter 2 and Chapter 3, we introduced two most important phenomena of our topic studied: phase transformation and damage, as a review and conclusion of current situation of these scientific fields. The knowledge in these two chapters is a foundation of further study on damage in multiphase metallic materials, which will be studied and explained in Chapter 4.

Chapter 4 Evolution Damage Multiphase Modelling

4.1 Introduction

The objective of this chapter is to develope a multiphase damage thermal viscoelastoplastic model. Compared to the constitutive equations in section 3.3, an additional variable z is introduced in this chapter. Here, volume fraction of the phase z_i is a function of temperature because the phase transformation is supposed to be driven by thermal loading. Mechanics coupled with damage depends on phase fraction and phase transformation. Therefore, constitutive equations in multiphase are coupled with damage, temperature and metallurgy. Two models will be developed and introduced in the chapter: mesoscopic damage model and two-scale damage model.

The section is limited to a 2-phase model, and we make a convention that 1 represents martensitic phase and 2 refers austenite. In terms of martensitic stainless steel, two phase transformations are included: $1 \rightarrow 2$ and $2 \rightarrow 1$. Volume fraction of martensite is z_1 and volume fraction of austenite is z_2. We have: $z_1 + z_2 = 1$.

Before developing constitutive equations, several hypotheses should be given:

Hypothesis 1: the materials are initially isotropic.

Hypothesis 2: the damage is isotropic.

Hypothesis 3: the damage has no influence on metallurgy.

Hypothesis 4: we consider additive decomposition of total strain in a thermoelastic part and plastic one. There is no uncoupling between elasticity and plasticity.

Hypothesis 5: the viscoplastic strain occers only when applied strain is large enough. The threshold surface is given as following:

$$f = J_2(\underline{\sigma} - \underline{X}) - R - \sigma_y \tag{4-1}$$

Hypothesis 6: the plasticity is supposed to be isotropic.

Let us recall some elementary notions for hardening plasticity. The plastic deformation of a material almost always causes hardening. At microscopic scale, hardening is characterized by an increase in the number of dislocations within material. With increasing dislocation density, they lose their mobilities. To allow accounting for this hardening, one defines a field of elasticity and studies its evolution according to the deformation. The evolution of this field can be represented by a combination of dilation, rotation, translation and distortion. Isotropic hardening is characterized by a homothetic swelling of the yield surface. In absence of distortion and on the basis of a circular surface (isotropy), isotropic hardening is then represented by the couple of scalars (r, R). Kinematic hardening is characterized by the translation of the center of the yield surface represented by a couple of tensors of internal variables $(\underline{\alpha}, \underline{X})$. Figure 4.1 shows these two hardenings when no damage occurs. The representation of hardenings with damage is given in Figure 4.2.

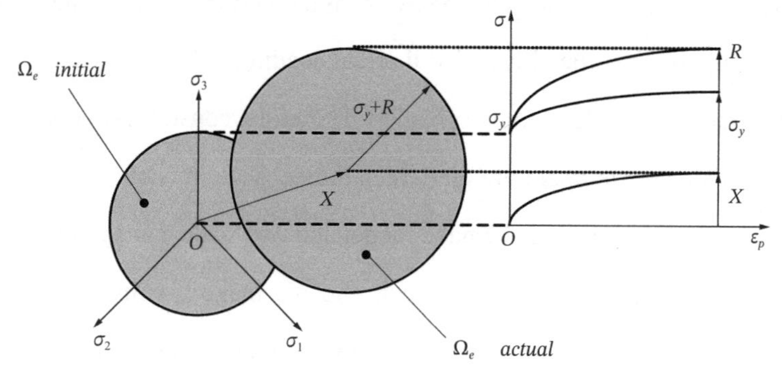

Figure 4.1 Geometrical representation of isotropic and kinematic hardenings

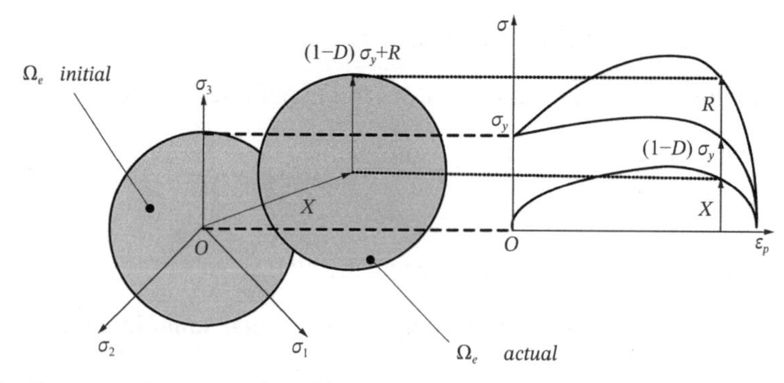

Figure 4.2 Geometrical representation of isotropic and kinematic hardenings coupling with damage

4.2 Definition of damage in multiphase

Because of assumption of isotopic and homogeneous damage, the RVE of two phases can be presented in 2-D. We suppose that each phase has its own damage. We distinguish a mesoscale for each damaged phase from a microscale to defined damage in each scale. The damage definition in two phases is shown in Figure 4.3.

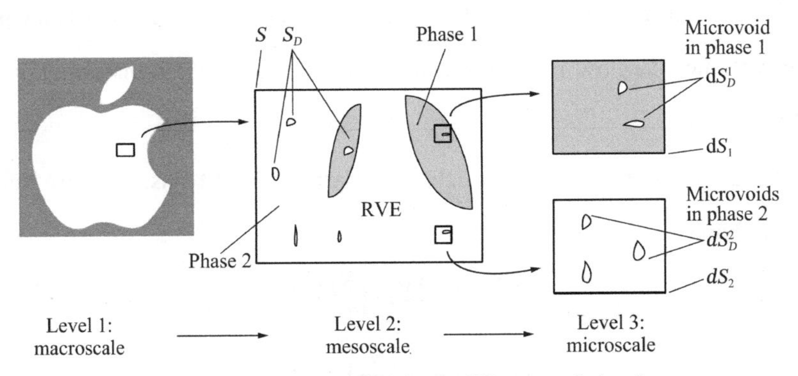

Figure 4.3 Damage definition at different scale levels

At mesoscopic scale, we neglect the fact the RVE consists of martensite (phase 1) and austenite (phase 2) and do not distinguish damage in the two phases. The damage variable is defined as the surface density of microvoides and microcracks.

$$D = \frac{S_D}{S} \tag{4-2}$$

At microscopic scale, the local damages can occur in martensite and austenite. Accordingly, the damage in phase i at microscale can be defined by:

$$D_i^\mu = \frac{\mathrm{d}S_D^i}{\mathrm{d}S_i}\ (i = 1{,}2) \tag{4-3}$$

The damage variable D in Equ. 4-2 can be divided into two parts: damage in phase 1 and damage in phase 2.

$$D = \frac{S_D}{S} = \frac{S_D^1}{S} + \frac{S_D^2}{S} = \frac{1}{S}\int_{s^1} \mathrm{d}S_D^1 + \frac{1}{S}\int_{s^2} \mathrm{d}S_D^2 \tag{4-4}$$

Introduce Equ. 4-3 into Equ. 4-4, we get:

$$D = \frac{1}{S}\int_{s'} D_1^\mu\,\mathrm{d}S_1 + \frac{1}{S}\int_{s^2} D_2^\mu\,\mathrm{d}S_2 \tag{4-5}$$

Let us define

$$D_i = \frac{1}{S_i}\int_{S_i} D_i^\mu\,\mathrm{d}S\ \ (i = 1{,}2) \tag{4-6}$$

$$\xi_i = \frac{S_i}{S}\ (i = 1{,}2\quad \xi_1 + \xi_2 = 1) \tag{4-7}$$

where D_i represents average damage in phase i; ξ_i is damaged surface fraction.

We get

$$D = D_1\xi_1 + D_2\xi_2 \tag{4-8}$$

The damage variable is defined and deduced from surface fraction whereas the phase fraction is defined by volume. We now introduce volume fraction in phase transformation into damage constitutive equations.

Under the assumption that the particles of martensite and austenite are spherical shape (Figure 4.4), we have:

$$z_1 = \frac{V_1}{V} = \left(\frac{R_1}{R_2}\right)^3 \tag{4-9}$$

$$\xi_1 = \frac{S_1}{S} = \left(\frac{R_1}{R_2}\right)^2 = z_1^{2/3} \tag{4-10}$$

$$z_2 = 1 - z_1 \tag{4-11}$$

$$\xi_2 = 1 - \xi_1 \tag{4-12}$$

Therefore, we get:

$$D = D_1\xi_1 + D_2\xi_2 = D_1 z_1^{2/3} + D_2\left(1 - z_1^{2/3}\right) \tag{4-13}$$

The damage mixture law is deriven by volume fraction at power 2/3.

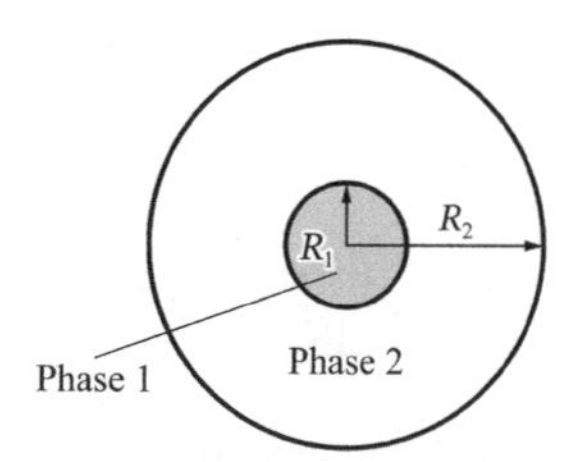

Figure 4.4 Schematic illustration of two phases with spherical shape

4.3 A proposed damage equation

Based on the study of Lemaitre-Chaboche damage model and Borona's equation, a damage strain model is proposed in the framework of continuum damage mechanics in this section.

The Lemaitre-Chaboche damage model in section 3.4.3 can be written as following in the uniaxial cases:

$$D = \frac{D_c}{p_R - p_D}\langle p - p_D\rangle \tag{4-14}$$

In the above equation, the damage is linear with plastic strain. But it is not in good agreement with some materials' characteristics. For example, the strain-damage law does not fit the nonlinearty of 15-5PH tensile test at room temperature (see Figure 5.20 for damage experimental results of 15-5PH).

Borona's damage equation adopted the exponent form of plastic strain, and the damage exponent α was successfully identified through experimental method[18]. However his model is coupled with an elastoplastic Ramberg-Osgood power law which is not very adapted to the material like 15-5PH studied here.

We propose now an extension of Lemaitre-Chaboche damage model with multiaxial strain.

The damage dissipation potential is chosen as:

$$F_D = -\frac{1}{s+1}\left(\frac{Y}{S}\right)^{s+1} \kappa \cdot \langle p - p_D \rangle^{\kappa-1} \tag{4-15}$$

where S, s and κ are material parameters which depend on temperature.

$$\dot{D} = -\frac{\dot{\lambda}(\partial F_D)}{\partial Y} = -\left(\frac{Y}{S}\right)^{s} \kappa \cdot \langle p - p_D \rangle^{\kappa-1} \cdot \dot{p} \quad \text{with } \dot{\lambda} = \dot{p} \tag{4-16}$$

with

$$-Y = \frac{\sigma_{eq}^2}{2E(1-D)^2} f\left(\frac{\sigma_H}{\sigma_{eq}}\right) \tag{4-17}$$

$$f\left(\frac{\sigma_H}{\sigma_{eq}}\right) = \left[\frac{2}{3}(1-\upsilon) + 3(1-2\upsilon)\left(\frac{\sigma_H}{\sigma_{eq}}\right)\right] \tag{4-18}$$

The ductile damage occurs only when the strain-hardening is saturated and the material is then considered perfectly plastic. The expression of the plastic criterion $\sigma_{eq}/(1-D) - R - k = 0$ shows that:

$$\frac{\sigma_{eq}}{1-D} = \tilde{\sigma}_{eq} = \text{constant} = K \tag{4-19}$$

Then

$$\dot{D} = \left[\left(\frac{K^2}{2ES}\right) \cdot f\left(\frac{\sigma_H}{\sigma_{eq}}\right)\right]^{s} \kappa \cdot \langle p - p_D \rangle^{\kappa-1} \cdot \dot{p} \tag{4-20}$$

Integrating from t_0 to $t\left(p|_{t_0} = p_D, p|_t = p, D|_t = D\right)$, we have:

$$D = \left[\left(\frac{K^2}{2ES}\right) \cdot f\left(\frac{\sigma_H}{\sigma_{eq}}\right)\right]^{s} \langle p - p_D \rangle^{\kappa} \tag{4-21}$$

At time t_c ($p|_{t_c} = p_R$, $D|_{t_c} = D_c$), we get the critical damage as:

$$D_c = \left[\left(\frac{K^2}{2ES} \right) \cdot f\left(\frac{\sigma_H}{\sigma_{eq}} \right) \right]^S \langle p_R - p_D \rangle^\kappa \tag{4-22}$$

In proportional loading cases, σ_H/σ_{eq} is a constant. Consequently, from the above two equations, we get:

$$D = D_c \left\langle \frac{p - p_D}{p_R - p_D} \right\rangle^\kappa \tag{4-23}$$

with

$$p_R - p_D = D_c^{1/\kappa} \left[\left(\frac{K^2}{2ES} \right) \cdot f\left(\frac{\sigma_H}{\sigma_{eq}} \right) \right]^{-s/\kappa} \tag{4-24}$$

If we take the initial damage D_0 of material into account, the proposed equations are written as:

In uniaxial loading:

$$D = \frac{D_c - D_0}{(p_R - p_D)^\kappa} \langle p - p_D \rangle^\kappa \tag{4-25}$$

In multiaxial loading:

$$D = \frac{D_c - D_0}{(p_R - p_D)^\kappa} \left\langle p \left[\frac{2}{3}(1+\upsilon) + 3(1-2\upsilon)\left(\frac{\sigma_H}{\sigma_{eq}} \right)^2 \right] - p_D \right\rangle^\kappa \tag{4-26}$$

The damage exponent κ is a parameter of material depending on the temperature.

In fact, if we neglect the initial damage of material and κ is equal to 1 , the proposed model is simplified into Lemaitre's damage model. Compared with Lemaitre model, the constitutive equation of the proposed damage model contains two additional parameters: D_0 and κ. It includes 5 material parameters

$$P = (p_D,\ \ p_R,\ \ \kappa,\ \ D_0,\ \ D_c) \tag{4-27}$$

Identification of this model consists of the quantitative evaluation of the five coefficient characteristics by mean of tensile tests.

4.4 Constitutive equations of mesoscopic model in multiphase

4.4.1 State potential

In this section, we introduce a mesoscopic model to describe damage under phase transformation. This damage is defined in mesoscale (Figure 4.3, Equ.4-4), i.e. there is no difference between the damage in austenite and in martensite.

The state potential of each phase is introduced through the method of local state in the thermodynamics of irreversible process. The total internal energy or the Helmholtz free energy is the sum of each phase's energy based on the linear mixture of different phases. In our study, it is restricted in isothermal process, and the totally internal energy e^{tot} consists of elastic and thermometallurgic internal energy $e^{e-th-met}$, viscoplastic (including isotropic and kinematic hardening) internal energy e^{vpi-pc} and internal energy induced by phase transformation e^{tr}. The following equations show these hypotheses:

$$e^{tot}(\underline{\varepsilon}^e, T, p, \underline{\alpha}, D, z_i) = e^{e-th-met}(\underline{\varepsilon}^e, T, D, z_i) + e^{vpi-vpc}(\varepsilon^e, T, p, \underline{\alpha}, D, z_i) + e^{tr}(T, z) \tag{4-28}$$

with:

$$e^{e-th-met}(\underline{\varepsilon}^e, T, D, z_i) = \sum_{i=1,2} \rho_i \psi_i^{e-th}(\varepsilon^e, T, D, z_i) + \rho\psi^{met}(\varepsilon^e, T, z_i) \tag{4-29}$$

$$e^{vpi-vpc}(\underline{\varepsilon}^e, T, p, \underline{\alpha}, D, z_i) = \sum_{i=1,2} \rho_i \psi_i^{vpi}(p, T, D, z_i) + \sum_{i=1,2} \rho_i \psi_i^{vpk}(\underline{\alpha}, T, D, z_i) \tag{4-30}$$

$$e^{tr}(T, z) = \rho\psi^{tr}(T, z) \tag{4-31}$$

where

$\rho_i \psi_i^{e-th}$: thermoelastic energy of phase i

$\rho\psi^{met}$: metallurgic energy during transformation

$\rho_i \psi_i^{vpi}$: blocked energy induced by isotropic hardening of phase i

$\rho_i \psi_i^{vpk}$: blocked energy induced by kinematic hardening of phase i

$\rho\psi^{tr}$: phase-transformation energy

And $\rho_i \psi_i^{e-th}$ can be defined by:

$$\rho_i \psi_i^{e-th}(\underline{\varepsilon}^e, T, D, z_i) = z_i \Big\{ \frac{1}{2}(1-D) A_i(T) \underline{\varepsilon}^e : \underline{\varepsilon}^e - \left[\alpha_i^{Tref}(T)(T - T_{ref}) - \alpha_i^{Tref}(T_0)(T_0 - T_{ref})\right] A_i(T) I : \underline{\varepsilon}^e \Big\} - z_i C_i (T - T_0)^2 \tag{4-32}$$

$$\rho\psi^{met}(\underline{\varepsilon}^e, T, z_i) = (z_2^0 - z_2) \Delta\underline{\varepsilon}_{1-2} A_2(T) I : \underline{\varepsilon}^e \tag{4-33}$$

The blocked energy by the mechanism of the isotropic hardening coupled with the temperature is selected in exponential form:

$$\rho_i \psi_i^{vpi}(p,T,D,z_i) = z_i c_i (1-D)\left[p + \frac{1}{\gamma_i}\exp(-\gamma_i p)\right] \tag{4-34}$$

where the blocked energy by the mechanism of kinematic hardening coupled with the temperature is described a quadratic form of $\underline{\alpha}$.

$$\rho_i \psi_i^{vpk}(\underline{\alpha},T,D,z_i) = z_i \frac{1}{3} g_i(T) b_i(T)(1-D)\underline{\alpha}:\underline{\alpha} \tag{4-35}$$

We propose to take the $\rho\psi^{tr}(T,z)$ as a simple expression:

$$\rho\psi^{tr}(T,z) = (1-z_\gamma)\left\langle \frac{\dot{z}}{|\dot{z}|}\right\rangle L(T) \tag{4-36}$$

where L represents the latent heat of phase transformation which can, per assumption, be taken as constant with the temperature.

The partial differentials of the free energy provide the following state's laws:

$$\underline{\sigma} = \frac{\partial e^{tot}}{\partial \underline{\varepsilon}^e} = \sum_{i=1,2} z_i \{(1-D)A_i(T)\underline{\varepsilon}^e - \left[\alpha_i^{Tref}(T)(T-T_{ref}) - \alpha_i^{Tref}(T_0)(T_0-T_{ref})\right]A_i(T)I\} + A_2(T)(z_2^0 - z_2)\Delta\underline{\varepsilon}_{1-2} \tag{4-37}$$

$$R = \frac{\partial e^{tot}}{\partial p} = \sum_{i=1,2} z_2 c_i(T)(1-D)[1-\exp(-\gamma_i p)] \tag{4-38}$$

$$\underline{X} = \frac{\partial e^{tot}}{\partial \underline{\alpha}} = \sum_{i=1,2} z_i \frac{2}{3} g_i(T) b_i (1-D)\underline{\alpha} \tag{4-39}$$

$$Y = \sum_{i=1,2} z_i \left[\frac{1}{2}A_i(T) + \underline{\varepsilon}^e \frac{1}{3} g_i(T)\underline{\alpha}:\underline{\alpha} + c_i(T)p\right] \tag{4-40}$$

$$s = -\frac{\partial e^{tot}}{\partial T} \tag{4-41}$$

$$k = -\sum_{i=1,2} \frac{\partial e^{tot}}{\partial z_i} \tag{4-42}$$

4.4.2 Dissipation potential

The yield function of each phase can be written as:

$$f_i = f_i(\underline{\sigma},R,\underline{X},D) = \frac{J_2(\underline{\sigma}-\underline{X})}{1-D} - \frac{R}{1-D} - \sigma_i^y \tag{4-43}$$

then

$$F_i = f_i + \frac{a_i}{2(1-D)} J_2^2(\underline{X}) + \frac{b_i}{2(1-D)} R^2 + \frac{S_i(T)}{[s_i(T)+1](1-D)} \left[-\frac{Y}{S_i(T)} \right]^{s_i(T)+1} \tag{4-44}$$

where

σ_{yi} : the yield stress of phase i

The pseudo-potential of dissipation φ is the function of all the dual variables:

$$\varphi = \varphi\left(\underline{\sigma}, A_k, \overrightarrow{grad}T, Y, k; \underline{\varepsilon}^e, V_k, T, D, z\right) \tag{4-45}$$

The total potential of dissipation is written by the sum of that of each phase:

$$\varphi = \sum_{i=1,2} \left[\varphi_i^{vp} + \varphi_{1-2}^{ptr} + \varphi_i^{th} + \varphi_i^{tr} + \varphi_i^{d} \right] \tag{4-46}$$

φ_i^{vp} : correspond to the classical viscoplastic potential, which is defined:

$$\varphi_i^{vp} = z_i \frac{u_i}{1+n_i} \langle f_i \rangle^{1+n_i} \tag{4-47}$$

φ_{1-2}^{ptr} : potential of phase-transformation, which is an only function of the stress because, at the temperatures of phase change, strain hardening effects are negligible.

$$\varphi_{1-2}^{ptr} = \frac{3}{4} \sigma^D : \sigma^D F_{1-2}'(z_{1-2}) \langle \dot{z}_{1-2} \rangle \tag{4-48}$$

φ_i^{tr} : potential metallurgical transformation, which is a transformation by diffusion, a temporal variation of the temperature influences the speed of diffusion considerably, by consequence the kinetics of transformation. It is defined by:

$$\varphi_i^{tr} = G_i\left(T, \underline{\varepsilon}^p, z_i\right) \langle -z_i \rangle \tag{4-49}$$

with

$G\left(T, \underline{\varepsilon}^p, z\right)$: a function to be determined phenomenologically.

To ensure the positivity of this potential, we propose to take the function to be a Heaviside step function[3]:

$$H(k) = \langle -k \rangle \tag{4-50}$$

φ_i^d the damage potential which is defined as:

$$\varphi_i^d = \frac{z_i S_i}{(s_i+1)(1-D)} \left(\frac{-Y}{S_i} \right)^{s_i+1} \tag{4-51}$$

φ_i^{th} the thermal potential is written:

3 The Heaviside step function, H, also called unit step function, is a discontinuous function whose value is zero for negative argument and one for positive argument.

$$\varphi_i^{th} = z_i \frac{1}{2} K_i(T) \left(\frac{grad(T)}{T} \right)^2 \tag{4-52}$$

The whole of the selected variables and the potentials previously defined lead to a thermodynamically acceptable model whose general formulation is presented here.

The viscoplastic flow:

$$\underline{\dot{\varepsilon}}^p = \frac{\partial \varphi}{\partial \underline{\sigma}} = \frac{3}{2} \frac{\dot{p}}{(1-D)} \frac{\underline{\sigma}^D - \underline{X}^D}{J_2(\underline{\sigma} - \underline{X})} + \frac{3}{2} \underline{\sigma}^D \sum_{i=1,2} F_i'(z_i) \langle \dot{z}_i \rangle \tag{4-53}$$

The evolution of the internal variable associated with isotropic hardening:

$$\dot{p} = \frac{\partial \varphi}{\partial R} = \sum_{i=1,2} z_i u_i \langle F_i \rangle^{n_i} \tag{4-54}$$

The evolution of the internal variable associated with isotropic hardening:

$$\underline{\dot{\alpha}} = \frac{\partial \varphi}{\partial \underline{X}} = \frac{3}{2} \frac{\dot{p}}{(1-D)} \frac{\underline{\sigma}^D - \underline{X}^D}{J_2(\underline{\sigma} - \underline{X})} - \sum_{i=1,2} u_i \langle F_i \rangle^{n_i} \underline{X} \tag{4-55}$$

The evolution of the damage variable:

$$\dot{D} = -\frac{\partial \varphi}{\partial Y} = \sum_{i=1,2} \frac{z_i S_i}{(1-D)} \left(\frac{-Y}{S_i} \right)^{s_i} \tag{4-56}$$

The kinetics of phase transformation:

$$\dot{z} = -\frac{\partial \varphi^{tr}}{\partial k} \tag{4-57}$$

In conclusion, the mesoscopic model takes into account, viscoplasticity with isotropic and kinematic hardening, temperature, phase transformation and damage. The phase transformation is introduced in the model by phase fraction dependent on temperature. The totally internal energy is the energy in each phase (by linear mixture of volume fraction) plus one induced by transformation plasticity.

4.5 Constitutive equations of two-scale multiphase model

4.5.1 Introduction

The two-scale model, which we introduce herein, is developed through a method of localizationhomogenization. The homogenizing procedure used is the Taylor's approximation[183], which assumes homogeneous deformations in a heterogeneous medium with nonlinear behavior. This law provides the closest possible match with Leblond's theoretical case for elastoplastic phases. Such an approach, called micro-macro, consists of starting from the behavior of each phase and working back to the macroscopic behavior of the material. After localization, the behaviors of each phase can be treated respectively, without coupling. Thus, the model can provide the freedom to choose the behavior type of each phase. Since each phase has its own mechanical behaviour, and the damage variable D_i in two phases is different: $D_1 \neq D_2$. The damage mechanics of phase depend on the material characteristics, which is related to temperature. The following hypotheses are introduced into the two-scale model.

Hypothesis 1: the rate of total macroscopic strain can be divided into two parts: the rate of microscopic strain of the phases and the transformation plasticity rate.

Hypothesis 2: there is no direct coupling between elastoplasticity and transformation plasticity.

Hypothesis 3: the microscopic strain rate of martensite is equal to the austenite strain rate in micro scale.

Hypothesis 4: there is no interaction between two phases, e.g. no damage or strain transfers from one phase to another.

Hypothesis 5: the homogenized macroscopic stress is obtained by the mixture of microscopic stress of each phase according to the volume fraction of the phases.

Hypothesis 6: we neglect the latent heat induced by phase transformation and the temperature in each phase keeps the same.

4.5.2 Strain localization

The approach of localization is based on the Taylor's approximation with equal repartition of strain rates in all phases of the multiphase composites. The classic strain rate at macroscale is equal to total strain rate of single phase at microscale. Thus, we get:

$$\dot{E}^c = \dot{\varepsilon}_i (i = 1,2) \tag{4-58}$$

According to the principle of localization mentioned above, we split the total strain ratio into two parts, one coming from the total microscopic strain rate of the phases, and the other representing the plastic transformation strain rate.

$$\dot{E}^{tot} = \dot{E}^{c} + \dot{E}^{pt} \tag{4-59}$$

$$\dot{E}^{tot} = \dot{\varepsilon}_i + \dot{E}^{pt} \forall i \quad \text{with } \dot{\varepsilon}_1 = \dot{\varepsilon}_2 \tag{4-60}$$

$$\dot{\varepsilon}_1 = \dot{\varepsilon}_i^{e} + \dot{\varepsilon}_i^{thm} + \dot{\varepsilon}_i^{vp} \forall i \tag{4-61}$$

The transformation plasticity strain rate is guided by Leblond's transformation plasticity model, which was introduced in section 2.6.2. A simplified form is as following equation:

$$\dot{E}^{pt} = \begin{cases} 0 & if \quad z \leqslant 0.03 \\ -\dfrac{3\Delta\mathcal{E}_{\gamma-a}^{th}}{\sigma_{\gamma}^{y}} \cdot S \cdot (\ln z) \cdot \dot{z} & if \quad z > 0.03 \end{cases} \tag{4-62}$$

After localization of strain, the problem of multiphase becomes the single phase behavior.

4.5.3 Mechanics in martensite

In terms of martensite, we choose the elastoplastic model with isotropic hardening. The constitutive equations are coupled with damage variable of martensite D_1.

The free energy of martensitic phase e_1^{lot} consists of two parts when the elastic behavior is uncoupled from the plastic behavior:

$$e_1^{tot} = e_1^{e}(T, \underline{\varepsilon}_1^{e}, D_1) + e_1^{P}(T, p_1, D_1) \tag{4-63}$$

The blocked energy by the mechanism of isotropic hardening coupled with the temperature is given as:

$$e_1^{P}(T, p_1, D_1) = \frac{a_1}{2} p_1^2 (1 - D_1) + b_1 p_1 \tag{4-64}$$

The thermoelastic free energy is:

$$e_1^{e}(T, \underline{\varepsilon}_1^{e}, D_1) = \frac{1}{2}(1 - D_1)\underline{\varepsilon}_1^{e} : A_1(T) : \underline{\varepsilon}_1^{e} + B_1(T) : \underline{\varepsilon}_1^{e} + C_1(T)(T - T_0)^2 \tag{4-65}$$

Therefore, the equations link between state variables and associated variables are:

$$\underline{\sigma}_1 = \frac{\partial e_1^{tot}}{\partial \underline{\varepsilon}_1^{e}} = (1 - D_1) A_1(T) : \underline{\varepsilon}_1^{e} + B_1(T) \tag{4-66}$$

$$R_1 = \frac{\partial e_1^{tot}}{\partial p_1} = a_1 p_1 (1 - D_1) + b_1 \tag{4-67}$$

$$Y_1 = -\frac{\partial e_1^{tot}}{\partial D_1} = Y_1^e + Y_1^p \tag{4-68}$$

$$Y_1^e = \frac{1}{2} A_1(T) \underline{\varepsilon}_1^e \tag{4-69}$$

$$Y_1^p = \frac{a_1}{2} p_1^2 \tag{4-70}$$

The pseudo-potential of dissipation is decoupled into the thermal dissipation and plastic dissipation, which is coupled with damage dissipation:

$$\varphi_1 = \varphi_1^p(\underline{\sigma}_1, R_1, D_1, T) + \varphi_1^{th}\left(\overrightarrow{gradT}, T\right) + \varphi_1^d(D_1, Y_1, T) \tag{4-71}$$

The thermal dissipation can be written as:

$$\varphi_1^{th}\left(\overrightarrow{gradT}, T\right) = \frac{1}{2} k_1(T) \frac{\overrightarrow{gradT}}{T} \cdot \frac{\overrightarrow{gradT}}{T} \tag{4-72}$$

The damage dissipation of martensite can be chosen as:

$$\varphi_1^d(D_1, Y_1, T) = -\frac{1}{2}\left(\frac{Y_1}{S_1}\right)^2 \kappa \cdot \langle p_1 - p_D \rangle^{\kappa - 1} \tag{4-73}$$

where S_1 and κ are material damage parameters of phase 1 depending on temperature. p_D is threshold plasticity to damage.

The yield function coupled with damage is given by:

$$f_1(\underline{\sigma}_1, R_1, D_1) = \frac{J_2(\underline{\sigma}_1)}{1 - D_1} - \frac{b_1 R_1}{1 - D_1} - \sigma_1^y(T) \tag{4-74}$$

The evolution laws are obtained from the generalized normality law:

$$\dot{\underline{\varepsilon}}_1^p = \dot{\lambda}_1 \frac{\partial \varphi_1}{\partial \underline{\sigma}_1} = \frac{3}{2} \frac{\underline{\sigma}_1^D}{J_2(\underline{\sigma}_1)} \frac{\dot{p}_2}{1 - D_1} \tag{4-75}$$

$$\dot{p}_1 = \dot{\lambda}_1 \frac{\partial \varphi_1}{\partial R_1} = \dot{\lambda}_1 \frac{b_1}{1 - D_1} \tag{4-76}$$

$$\dot{D}_1 = -\dot{\lambda}_1 \frac{\partial \varphi_1}{\partial Y_1} = \frac{Y_1}{S_1} \cdot \kappa \cdot \langle p_1 - p_D \rangle^{\kappa - 1} \cdot \dot{p}_1 \tag{4-77}$$

4.5.4 Mechanics in austenite

For the austenite, the elastoplastic model with linear kinematic work hardening is employed. The constitutive equations are coupled with damage variable of austenite D_2.

The free energy of austenite phase e_2^{tot} is divided into two parts when the elastic behavior is uncoupled from the plastic behavior:

$$e_2^{tot} = e_2^e(T, \underline{\varepsilon}_2^e, D_2) + e_2^p(T, \underline{\alpha}_2, D_2) \tag{4-78}$$

The blocked energy by the mechanism of kinematic hardening coupled with damage variable is described by a quadratic form of $\underline{\alpha}_2$:

$$e_2^p(T, \underline{\alpha}_2, D_2) = \frac{1}{3} g_2(T)(1 - D_2)\underline{\alpha}_2 : \underline{\alpha}_2 \tag{4-79}$$

The thermoelastic free energy is:

$$e_2^e(T, \underline{\varepsilon}_2^e, D_2) = \frac{1}{2}(1 - D_2)\underline{\varepsilon}_2^e : A_2(T) : \underline{\varepsilon}_2^e + B_2(T) : \underline{\varepsilon}_2^e + C_2(T)(T - T_0)^2 \tag{4-80}$$

Therefore, the equations link between state variables and associated variables are:

$$\underline{\sigma}_2 = \frac{\partial e_2^{tot}}{\partial \underline{\varepsilon}_2^e} = (1 - D_2)A_2(T) : \underline{\varepsilon}_2^e + B_2(T) \tag{4-81}$$

$$\underline{X}_2 = \frac{\partial e_2^{tot}}{\partial \underline{\alpha}_2} = \frac{2}{3} g_2(T)(1 - D_2)\underline{\alpha}_2 \tag{4-82}$$

$$Y_2 = -\frac{\partial e_2^{tot}}{\partial D_2} = Y_2^e + Y_2^p \tag{4-83}$$

$$Y_2^e = \frac{1}{2} A_2(T)\underline{\varepsilon}_2^e \tag{4-84}$$

$$Y_2^p = \frac{1}{3} g_2(T)\alpha_2 : \alpha_2 \tag{4-85}$$

The pseudo-potential of dissipation is decoupled into the thermal dissipation and plastic dissipation, which is coupled with damage dissipation:

$$\varphi_2 = \varphi_2^p(\underline{\sigma}_2, \underline{X}_2, D_2, T) + \varphi_2^{\varphi th}(\overrightarrow{gradT}, T) + \varphi_2^d(D_2, Y_2, T) \tag{4-86}$$

The thermal dissipation can be written as:

$$\varphi_2^{th}(\overrightarrow{gradT}, T) = \frac{1}{2} k_2(T) \frac{\overrightarrow{gradT}}{T} \cdot \frac{\overrightarrow{gradT}}{T} \tag{4-87}$$

The damage dissipation of austenite chosen φ_1^d is the exponential function:

$$\varphi_2^d(D_2, Y_2, T) = \frac{S_2}{(s_2 + 1)(1 - D_2)} \left(-\frac{Y_2}{S_2}\right)^{s_2 + 1} \tag{4-88}$$

where s_2 and S_2 are material damage parameters of phase 2 depending on temperature.

The yield function of austenite can be written:

$$f_2(\underline{\sigma_2}, \underline{X_2}, D_2) = \frac{J_2(\underline{\sigma_2} - \underline{X_2})}{1 - D_2} - \sigma_2^y(T) \quad (4\text{-}89)$$

The potentials previously defined lead to a thermodynamical model. The evolution laws are obtained from the generalized normality law:

$$\underline{\dot{\varepsilon}}_2^p = \dot{\lambda}_2 \frac{\partial \varphi_2}{\partial \underline{\sigma_2}} = \frac{3}{2} \frac{\underline{\sigma}_2^D - \underline{X}_2^D}{J_2(\underline{\sigma_2} - \underline{X_2})} \frac{\dot{p}_2}{1 - D_2} \quad (4\text{-}90)$$

$$\dot{\underline{\alpha}}_2 = \dot{\lambda}_2 \frac{\partial \varphi_2}{\partial \underline{X_2}} = \frac{3}{2} \frac{\dot{p}_2}{1 - D_2} \frac{\underline{\sigma}_2^D - \underline{X}_2^D}{J_2(\underline{\sigma_2} - \underline{X_2})} \quad (4\text{-}91)$$

$$\dot{D}_2 = -\dot{\lambda}_2 \frac{\partial \varphi_2}{\partial Y_2} = \frac{\dot{\lambda}_2}{1 - D_2} \left(-\frac{Y_2}{S_2}\right)^{s_2} \quad (4\text{-}92)$$

4.5.5 Memory effect during phase change

The memory effect describes that internal variables in daughter phase are inherited form mother phase when one phase (mother phase) disappears and the other phase (daughter phase) occurs. Memory coefficient determines how many percent values of internal variables transfer from an old phase to a new phase. Herein, η is memory coefficient of damage variable, μ_i is memory coefficient of internal variable A_i. The following equations give the relationship between internal variables in daughter phase and in mother phase:

$$D^{daughter} = \eta \cdot D^{mother} \quad (4\text{-}93)$$

$$A_i^{dughter} = \mu_i \cdot A_i^{mother} \quad (4\text{-}94)$$

In the equations, the memory effect is null if $\eta = 0$ or $\mu = 0$ and full if $\eta = 1$ or $\mu = 1$.

4.5.6 Stress and damage homogenization

The homogenized macroscopic stress is obtained by a linear law of mixture guided by the volume fraction of each phase. This simple homogenization could take stress at microscale back to macroscopic stress, which is involved in structure analyses.

$$\Sigma = \sum_{i=1,2} z_i \underline{\sigma_i} \quad (4\text{-}95)$$

The damage variable is defined and deduced from the surface fraction whereas the phase fraction is defined by the volume. The damage homogenization is given by the following equation:

$$D = D_1\xi_1 + D_2\xi_2 = D_1 z_1^{2/3} + D_2\left(1 - z_1^{2/3}\right) \tag{4-96}$$

This two-scale model takes only two phases into account. It is not difficult to extend to situation with more phases. Compared to two-phase situation, multiphase has the same localization-homogenization laws. The difference lies in where the model for multiphase needs more identification of phases.

4.5.7 An example in one dimension

In this section, a simple example in one dimension is given to illustrate the two-scale model presented previously. The simulation was implemented in Matlab 6.5. We introduced a similar "Satoh"[4] type test in which a uniaxial bar is clamped at the top and bottom and simultaneously suffers thermal loading. The test bar is supposed to be cooled homogeneously from high temperature to room temperature. Thus, longitudinal displacements of bar are restrained during cooling. Thus, it may explain the same or similar phenomena as that can be observed in welding heat-affected zone (HAZ). The test contains several complex coupled phenomena, and they are thermal (temperature), metallurgical (phase-transformation) and mechanical (strain-stress, damage) behaviors.

We supposed the material properties as following:

Thermal strain difference between the two phases: $\Delta\varepsilon_{\alpha-\gamma}^{20°C} = 0.011$

Thermal expansion coefficient of the austenitic phase ($T_{ref} = 1000°C$):

$$\alpha_\gamma = 22.6 \times 10^{-6} + 2.52 \times 10^{-9} \cdot T$$

Thermal expansion coefficient of the martensitic phase:

$$\alpha_\alpha = 12.35 \times 10^{-6} + 7.710^{-9} \cdot T(T \leqslant 350°C); \alpha_\alpha = 15 \times 10^{-6}(350°C < T < 700°C)$$

The beginning temperature of martensitic transformation: $M_s = 400°C$

We suppose that the bar has the elasticity and perfect plasticity and that the Young's modulus

4 The test was performed by induction heating of a tensile specimen with both ends clamped. During the test, the reaction force-versus-temperature curves were recorded. The original Satoh test was designed for two purposes. First, for a given heat treatment, the Satoh test can be used for characterization of materials. Secondly, for a given material, the test can be used to study the effect of heat treatment.

and yield stress depend on temperature, which is given in Table 4.1. The damage is supposed only on martensite whereas there is no damage in austenite. The damage model was presented in Equ. 4-22 and Equ.4-23 in Section 4.3. Parameters of the damage model are supposed as: $R = 0.02$, $s = 6$, $p_D = 0$ and $\kappa = 1$. The temperature loading of the bar is from 1000°C to 20°C with $\dot{T} = -9.8$°C/s. The phase-transformation model adopts Koistinen and Marburger model in Equ. 2-3 in Section 2.3.4 (Figure 4.5).

Material properties depend on temperature Table 4.1

T(°C)	0	100	200	400	600	700	800	900	1000
E(GPa)	208	204	200	180	135	80	50	32	30
σ_{ym}(MPa)	1200	1170	1100	980	680	350	100	50	20
σ_{ya}(MPa)	140	130	120	110	100	70	60	30	20

m = martensite, a = austenite

Strain variables are shown in Figure 4.6, in which the TRIP strain is dominating comparing to the others. Such big value is due to the large macroscopic stress at low temperature. Furthermore, the TRIP strain that comes from the damage-mechanical model is smaller a little than one gained from the usual strain-stress model (without damage) because of the decline of stress in damage condition. The thermal strain has an observable effect to the stress in "Satoh" test, but the metallurgical strain from austenite to martensite weakens this effect because its volume increases whereas the thermal strain is negative.

From the Figure 4.7, it is observed that both austenite and martensite yield at the beginning (martensite, near to 400°C). Generally, the austenitic component of stress is much lower than the martensite's because of the big difference of yield stresses. It is evident that damage decreases the effective stress, especially for martensitic component. Before martensitic phase-transformation, all variables of martensite are null and the mechanical behavior is entirely determined by the austenitic phase. With the development of phase transformation, the martensite plays more and more role not only in evolution of stress but also to the damage's growth. At the end of phase-transformation stage, the macroscopic stress increases slowly or even decreases because of damage's effect. Damage evolution is shown in Figure 4.8.

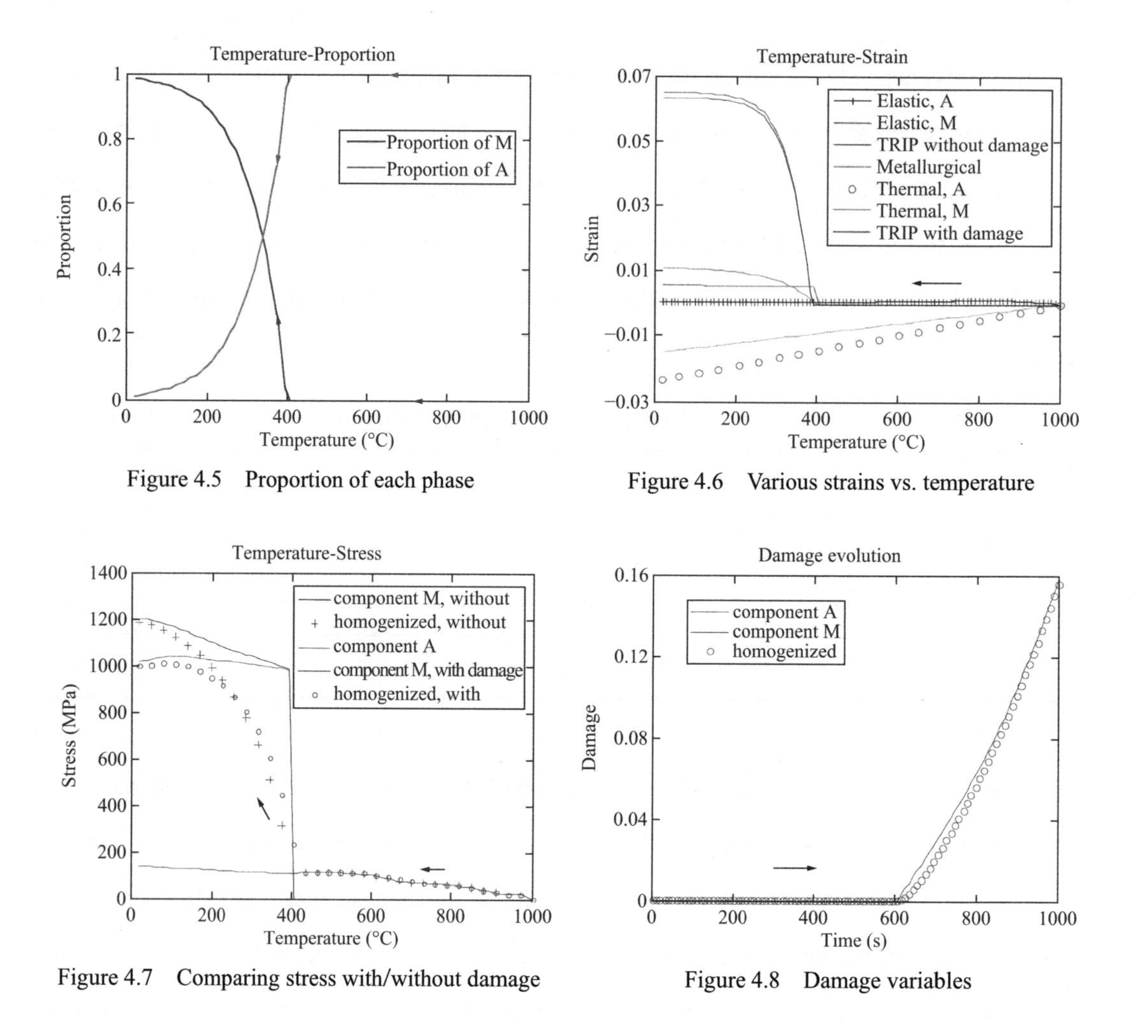

Figure 4.5 Proportion of each phase

Figure 4.6 Various strains vs. temperature

Figure 4.7 Comparing stress with/without damage

Figure 4.8 Damage variables

4.6 Conclusion

Two mechanical models coupled damage, thermal and metallurgy were presented in the chapter. The mesoscopic model was developed based on the linear mixture of dissipation in framework of thermodynamics and definition of damage variable at mesoscale. The benefit of mesoscopic model is simple calibration of damage parameters: neglecting the difference between each phase. The two-scale connects behaviors of multiphase and single behaviors through localization-homogenization laws. Such a modeling brings a great flexibility of calculation. On the one hand, each phase can have its own behavior (elastoplastic, viscoplastic...). On the other hand, one keeps a great freedom on the modeling of the plasticity of transformation.

Chapter 5 Experimental Study and Identification of Damage and Phase Transformation Models

5.1 Introduction

The constitutive equations were proposed in the Chapter 4. The parameters of transformation-plasticity model and damage model will be calibrated through experimental study in this chapter. On the one hand, phase transformation and transformation plasticity can be identified through tests of round bar with temperature loading. On the other hand, cyclical load-unload tests of round bar help for measurement of damage. In addition, we will introduce tensile tests of flat notched specimen with the help of digital image correlation system, in order to analyze and study strain localization and damage evolution on surface of specimen. In terms of flat notched specimens, we will not only compare that three notch sizes of specimen influence on strain localization and damage distribution, but also study the situation at different temperatures. At last, the microscopic observation will be implemented to study metallurgical and mechanical behaviours of the micro scale.

5.2 Design of specimens

Two categories of specimen are used in this study: round bar (RB) and flat notched specimen (FNS). The geometries of specimens are depicted in Figure 5.1. RB and FNS are adopted for the following purposes:

Tests of round bar provide strain-stress curve, dilatometry curves, mechanical properties, thermal behaviour, and damage variable D by measurement of the variation of elastic modulus.

Tests of flat notched specimen lead to the verification of damage model and strain localization due to the fact that various notch radii induce different triaxial stress ratio profiles in the mid section.

As far as tests of flat notched specimens are concerned, we designed three specimens with different radii of notch with $R = 1$mm (Case A), $R = 2.5$mm (Case B) and $R = 4$mm (Case C) because the notch geometry generates stress and strain gradients from the specimen centre to the notch edge. The notch influence on triaxiality ratio can be calculated with finite element method. Figure 5.2 gives an example of triaxiality distribution on surface of specimen with minimum diameter under elastic tension. One can clearly see triaxiality is maximum in the centre in Case C and near the notch for the Case A.

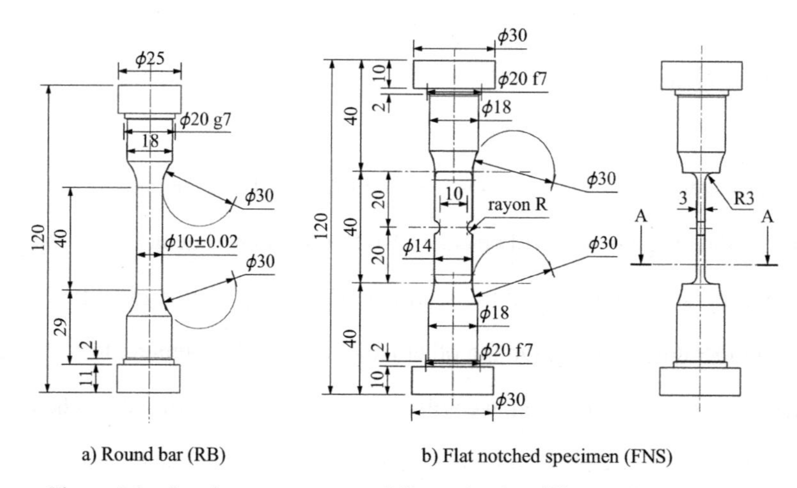

Figure 5.1 Specimen geometry and dimension in millimeter (RB and FNS)

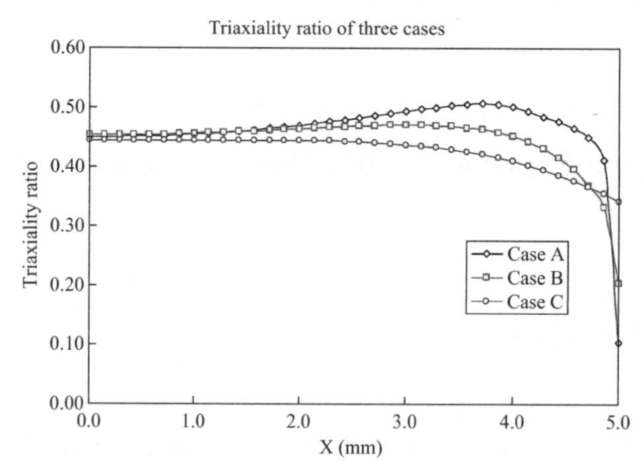

Figure 5.2 Notches (Case A, Case B and Case C) influence on triaxiality (Distribution along minimum section under elastic tension)

5.3 Experimental devices

The tests are performed on a MTS SCHENCK servo-electro-hydraulic tension-compression machine with maximum capacity of 250kN (Figure 5.3a). Heating is generated by electromagnetic induction, and power is supplied by a 3kW Celes generator (Figure 5.3d). Water-cooling setup provides fast cooling rate. An extensometer with the precision of 1 micron is used to measure the strain. It consists of two sticks, connecting components, fixing component and conducting wires. The two sticks are placed in parallel and the gauged distance is 15 millimeters. The thermal couples connect the Converter TEPI (Model BEP304) for signal amplification and conversion; these signals are then collected by the FlexTest control system (Figure 5.3c). The automated configuration includes the controller, a PC, and the model 793.00 system software bundle.

In the tests of flat notched specimen, two C.C.D. cameras (Figure 5.3b) are used to record digital images of the sample's surface. 3D digital image correlation (LIMESS measurement technique and software) are utilized to process digital images and analyze the strain localisations. The application of C.C.D cameras and image correlation makes it possible to monitor strain localization and observe ductile damage evolution.

a) MTS SCHENCK

b) General view

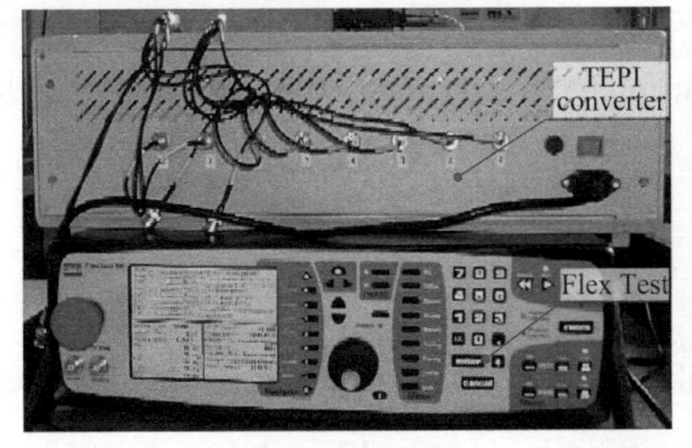

c) TEPI and FlexTest

d) Celes generator

Figure 5.3 Experimental setup

5.4 Measurement

5.4.1 Temperature

The temperature measurements are performed using thermocouples (type K) microwelded on the surface of the specimen (Figure 5.4). Four thermocouples (T1, T2, T3 and T4) are used to monitor the temperature in the useful zone. One thermocouple (T0) is connected to the Heat inductor CELES for temperature feedback and control.

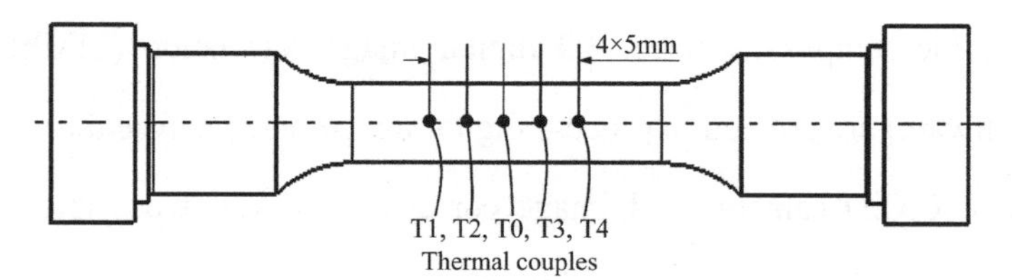

Figure 5.4 Thermal couples welded on surface of specimen

5.4.2 Force and displacement

Global force and displacement are respectively measured by the loading system: load cell and LVDT (Linear Variable Displacement Transducers) sensor of tensile test machine are recorded in Controller System (PC component).

5.4.3 Stress

The nominal stress σ^* is gained from force F and initial cross-sectional area S_0 by the formula:

$$\sigma^* = \frac{F}{S_0} \tag{5-1}$$

Considering the cross-section will shrink, the initial cross-sectional area should be replaced by the effective one. It is supposed the material is plastic incompressible and the volume keep the same in the different force loading condition. And then the nominal stress leads to true stress guided by:

$$\sigma = \frac{F}{S} = \frac{F}{S_0}\frac{L}{L_0} = \sigma^*\left(\frac{L_0 + \Delta L}{L_0}\right) = \sigma^*(1 + \varepsilon^*) \tag{5-2}$$

5.4.4 Strain

The local strain is measured by the extensometer (Figure 5.5). Such recorded strain is called nominal strain. In order to obtain the strain (Hencky strain), the following formula is applied:

$$\varepsilon = \int_{l_0}^{l}\frac{\mathrm{d}l}{l} = \ln\left(\frac{L}{L_0}\right) = \ln\left(\frac{\Delta L + L_0}{L_0}\right) = \ln(1 + \varepsilon^*) \tag{5-3}$$

where

L_0 : the initial length of measured part

ΔL : the elongation

ε^* : nominal strain

ε : true strain

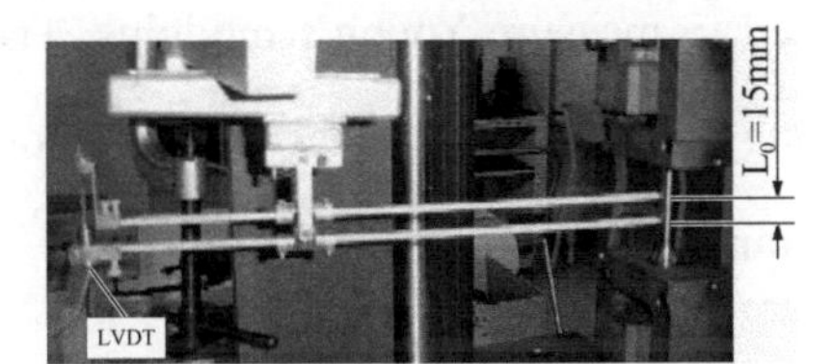

Figure 5.5 Extensometer

5.4.5 Damage

Damage is not easy to be measured directly. Its quantitative evaluation, like any physical value, is linked to the definition of the variable chosen to represent the phenomenon. We can measure damage by measuring the modification of the mechanical properties of elasticity. The method of variation of the modulus of elasticity, which is a non-direct measurement, is preferred because of its simple principle and great precision (Figure 3.2).

Recall effective stress in the case of uniaxial tension [124]:

$$\tilde{\sigma} = \frac{\sigma}{1 - D} = E\varepsilon_e \tag{5-4}$$

$E(1 - D) = \tilde{E}$ can be interpreted as the elastic modulus of the damaged material. Damage variable D can be written:

$$D = 1 - \tilde{E}/E \tag{5-5}$$

where E is Young's modulus at moment of undamaged material. $\tilde{E}$ represents Young's modulus at moment of damaged material.

Despite its apparent simplicity, this measurement is rather tricky for the following reasons:

1) the measurement of modulus of elasticity requires precise measurement of very small strains;

2) damage is usually localized, thus requires a very small extensometry base (about 0.5 to 5 mm);

3) the best straight line in the strain-stress graph representing elastic loading or unloading is difficult to define;

4) at high temperature, metal becomes weak and strain-stress curve shows very strong nonlinearity.

Therefore, an adequate procedure is chosen to optimize the measure of Young's modulus. The stage of unloading ramp is used to measure Young's modulus. The evaluation of the elasticity modulus during elastic unloading must avoid the zones of strong nonlinearity. This is achieved by performing the evaluation in a range of stress defined as follows:

$$\sigma_{min} < \sigma < \sigma_{max} \tag{5-6}$$

where $\sigma_{min} = 0.15\sigma(\varepsilon)$, and σ_{max}: is the upper stress limit of linear part.

In the strain-stress region defined above, we introduced the error rectangle (Figure 5.6), which consists of strain error and stress error, to ensure that the useful experimental data are within in the error tolerance. Strain error is due to the errors of various devices involved in measurement of strains. Similarly, stress error is related to force measuring instrument. The resulting computed error rectangle width is: for stain error = 0.00005 and for stress error = 0.5MPa.

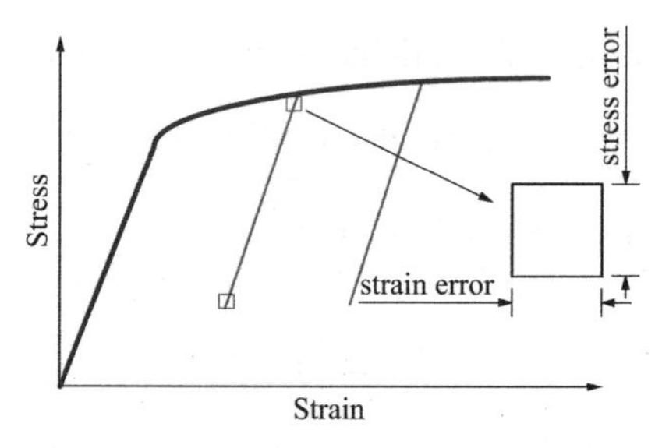

Figure 5.6 Error rectangle and linear measurement of strain-stress curve

5.5 Digital image correlation (DIC)

The digital image correlation was involved in our experimental study. An image processing gives the space-time evolution of various kinematic variables (displacement, velocity, strain, strain-rate, etc.) on the sample surface. The digital image correlation technique was originally introduced in the early 80s by researchers from the University of South Carolina [151]. The idea behind the method is to infer the displacement of the material under test by tracking the deformation of a random speckle pattern applied to the component's surface in digital images acquired during the loading. Mathematically, this is accomplished by finding the region in a deformed image that maximizes the normalized cross-correlation score with regard to a small subset of the image taken while no load was applied. By repeating this process for a large number of subsets, full-field deformation data can be obtained.

Helm et al. extended the DIC method to use multiple cameras, permitting the measurement of three-dimensional shape as well as the measurement of the three-dimensional deformation [93]. The three-dimensional technique requires the use of at least two synchronized cameras acquiring images of the loaded specimen from different viewing angles. By determining corresponding image locations across views from the different cameras and tracking the movement throughout the loading cycle, the shape and deformation can be reconstructed based on a simple camera calibration. Correlation techniques on digital images have been used to measure in-plane displacement components on plane samples by many researchers [37][204] and to study necking of sheet metal in the laboratory of contacts and solid mechanics of INSA-Lyon in France [23][41][40][185].

It is necessary for the acquisition of the qualified digital images. In order to obtain good digital image of sample during tensile test, we should pay attention to illumination, speckle pattern, choice of spray paint, calibration of two cameras and paint movement with the surface. Reflections and shadows should be absolutely avoided. Small speckles loads to high spatial resolution. Calibration is necessary for 3-D DIC, but no need for 2-D DIC. In our test, we used black and whiter spray paints to get proper speckle pattern and to coordinate the contrast gradient of image. However, it is difficult to find a spray paint with resistance to high temperature. The FNS with speckle pattern and FNS with image correlation are respectively shown in Figure 5.7a and Figure 5.7b.

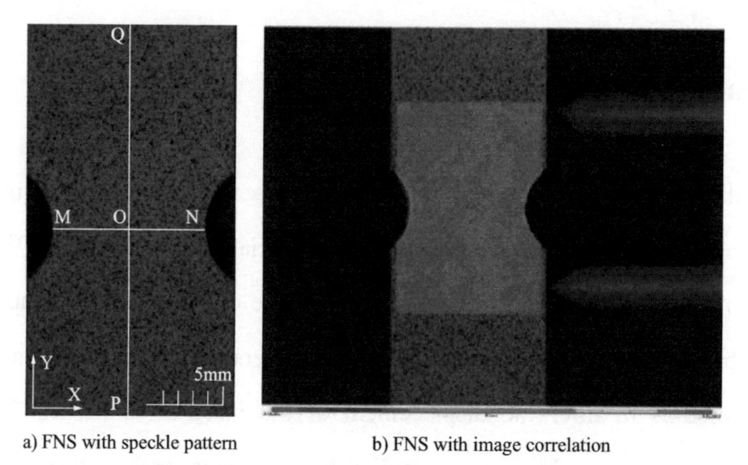

a) FNS with speckle pattern   b) FNS with image correlation

Figure 5.7 Flat notched specimen in tensile test

5.6 Experimental programs

5.6.1 Round bar tests

In terms of the tests of RB, three categories are designed:

Type 1: Free expansion-contraction tests, which aim to identify the thermal parameters, such as coefficient of heat expansion, M_s, A_{c_1}, A_{c_3} and the difference of thermal strain difference between two phases.

Type 2: Expansion-contraction tests with applied tensile load, which are used to study the phase transformation induced plasticity.

Type 3: Cyclic load-unload tests, which not only provide strain-stress curve and mechanic properties, such as Young's modulus, yield strain, yield stress and hardening behaviour, but also permit to identify the damage variable D.

For the test Type 1, one specimen with various thermal histories is implemented. The specimens are loaded with heating rate of 4-5°C/s and cooling rate of -3°C/s. It is imposed four thermal cycles: the first three cycles reach maximum temperature of 860°C, and the maximum temperature of last one is 1050°C (Figure 5.11).

The temperature history of the tests Type 2 (Figure 5.8) is similar with that of the test Type 1.

Specific experiments are designed to measure to identify the phase transformation induced plasticity through loading pressure at cooling stage. Loading pressures with value of 37.5MPa, 50MPa, 75MPa and 115.5MPa (Figure 5.8), are applied before M_S temperature during the cooling stage, and last to the completion of martensite transformation.

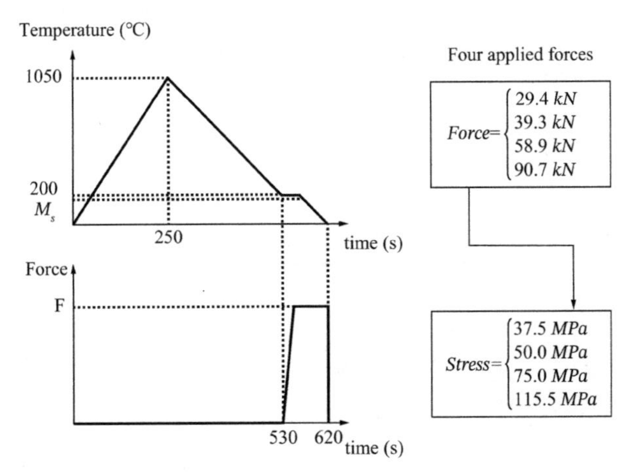

Figure 5.8 Temperature and force loads of TRIP tests (Test type 2)

As far as test Type 3 is concerned, eleven specimens (namely P_i, $i = 1,2,\cdots,11$) are used with different temperature history and stretched at several constant temperatures. There is no hold time at maximum temperature. The temperature heating rate is 4～5°C/s and the cooling one is about −2°C/s. The idea is to compare the stress strain response and the damage identification for a specimen at the same temperature having had a different temperature history. Table 5.1 is a summary of the performed experiments: one can observe that test P1 and P8 should produce the same material characteristic if the temperature history had no influence on material response. The tensile tests are displacement controlled. The displacement loading history consists of several cycles (loading and unloading), and up to rupture of specimens except the free expansion specimens. The detailed information of displacement loading is in Table 5.2. For example, the temperature and displacement loads of P6 are given in Figure 5.9.

Temperature loading condition of RB tests Table 5.1

Condition	P1	P2	P3	P4	P5	P6	P7	P8	P9	P10	P11
Test T (°C)	20	200	600	700	850	600	200	20	20	300	300
Max T (°C)	20	200	600	700	850	1050	1050	1050	860	860	860
Phase	M	M	M	M	A	A	A	M	M	A	A

Note: T presents temperature; M means martensite phase; A is austenite phase; heating rate: 4-5°C/s; cooling rate (average): −3°C/s.

Displacement loading condition of RB tests　　Table 5.2

	Condition	P1	P2	P3	P4	P5	P6	P7	P8	P9	P10	P11
Load	Disp. increment	1.0	0.8	0.8	0.6	0.8	0.8	1.0	1.0	1.0	0.8	0.8
	Time (s)	20	16	16	10	16	16	20	20	20	16	16
Unload	Disp. increment	−0.5	−0.3	−0.3	−0.5	−0.2	−0.2	−0.5	−0.5	−0.5	−0.2	−0.2
	Time (s)	10	6	6	1	4	4	10	10	10	4	4
Total Cycle		10	9	8	16	15	16	16	10	12	12	10

Note: unit of displacement is mm.

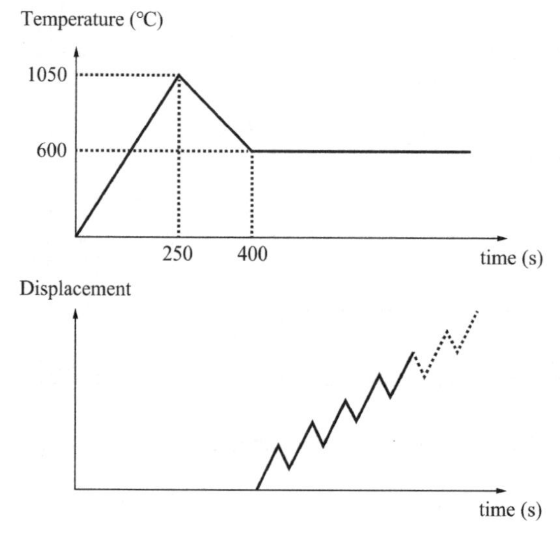

Figure 5.9　Temperature and displacement loads of P6 (at 600°C, Austenite)

5.6.2 Tensile tests of flat notched specimen

The tests of flat notched specimens with various notches were performed at various temperatures. The five conditions of test (namely F_i, $i = 1,2,\cdots,5$) were given in Table 5.3. In every condition, three samples (Case A, B, C) were tested. Consequence, fifteen specimens are used in this kind of test (namely F_{ij}, $i = 1,2,\cdots,5$, $j = A,B,C$, e.g. $F1A$ represents the specimen with thermal condition $F1$ and notch Case A).

FNS test conditions　　Table 5.3

Condition	$F1$	$F2$	$F3$	$F4$	$F5$
Test T (°C)	20	200	20	200	300
Heating rate (°C/s)	—	4-5	4-5	4-5	4-5
Cooling rate (°C/s)	—	—	−2	−2	−2
Max T (°C)	20	200	860	860	860
Phase	M	M	M	MIX	A

5.7 Experimental results and identification of parameters

5.7.1 Thermal and metallurgical data

The loaded temperature histories are measured through thermocouples welded on surface of specimen. For example, the temperatures of specimen P1 are recorded by thermocouples and the temperatures vs. time curves at four positions of specimen surface are shown in Figure 5.10. It shows the temperatures at middle of specimen have a good homogeneity, and the maximum difference $\Delta T_{max} = 14°C$.

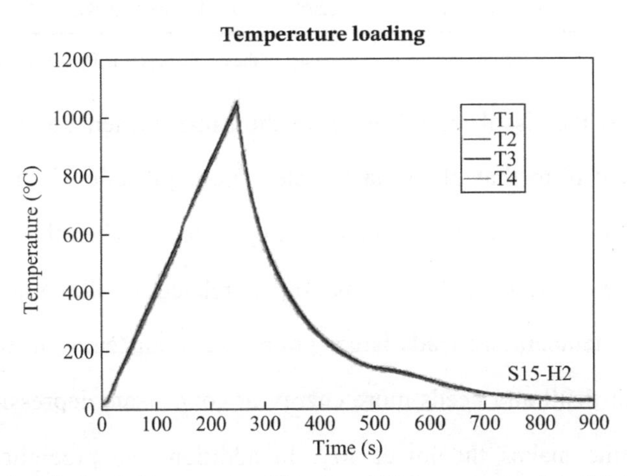

Figure 5.10 The loaded temperature histories of P1 (measured points: T1, T2, T3, T4)

Experiment results obtained from the tensile tests of round bar, such as coefficient of expansion, metallurgical characteristics are listed in the Appendix B. These basic data are necessary to simulate material behavior during heating and cooling (temperature history from 20°C to 1050°C) by linear interpolation between the indicated points.

In Figure 5.11, four cycles are imposed to the specimen. The first cycle is for the purpose of eliminating, at least partly, the internal stresses dependent on the means of rolling of sheet. The second and the third cycle are used for the thermo-metallurgical characteristics. The parameters of phase transformation are directly identified through fitting free expansion curves (Table 5.4).

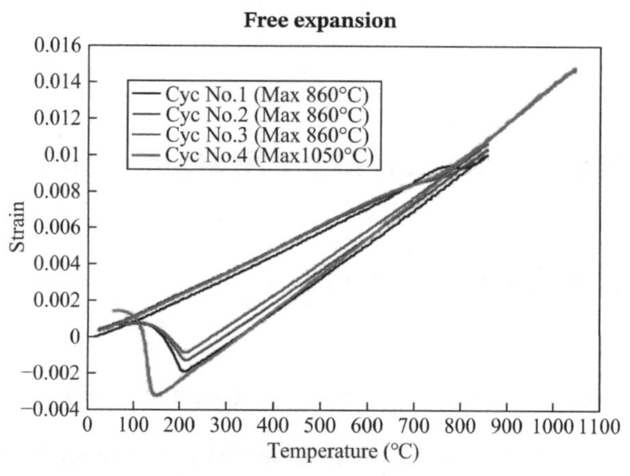

Figure 5.11 Free dilatometers with four cycles'thermal loading (the first three cycles reach maximum temperature of 860°C, whereas the maximum temperature of last one is 1050°C)

Parameters of phase transformation from experiments Table 5.4

Parameters	$M_s(1050)$	$M_s(860)$	A_{c_1}	A_{c_3}	α_α	α_γ	$\Delta\varepsilon_{\alpha-\gamma}^{th}$	β
Value	150°C	200°C	760°C	820°C	1.25E − 05	2.09E − 05	9.58E − 03	0.011

Note: $M_s(1050)/M_s(860)$ means that the maximum temperature of the thermal cycle is 1050/860°C.

It is noticed that the fist three cycles have the same martensite start temperature (M_S) whereas it is different from that of the last cycle. The explanation is based on the material metallurgy. Martensite is formed based on the gain of austenite when the temperature is cooled to M_S. Consequence, martensite transformation is related to the grain size of austenite. Generally, the higher temperature leads larger grain size of austenite, in which the martensite nucleation is more difficult and needs more energy or condensate depression. Thus, the larger grain size of austenite makes the lower M_S. In addition, the precipitation influences the transformation. At high temperature, the second phase is precipitated among grain boundaries, and lowers the martensite start temperature. Therefore, the grain size of austenite affects the martensite transformation kinematics. It should be taken into account in the simulation of hot processing.

5.7.2 Mechenical data

The mechanical behaviors of 15-5PH are identified by tensile test of round bar at various temperatures. The details of mechanical characteristics, such as Young's modulus, yield strength and ultimate strength, are given in the Appendix B. The Figure 5.12 and Figure 5.13 give the tensile curves of martensite and austenite at different temperatures. These data are necessary for welding simulation of 15-5PH.

An important phenomenon is observed in the tensile curves: the thermal histories have a large influence on mechanical characteristics. The maximum temperature has to be taken into account in the material studied here. In Figure 5.12, strain-stress curve for P1, P8 and P9 are plotted at room temperature; the curves have some largely important differences. P11 and P6 in Figure 5.13 are tested at 600°C, however P11 has much more strength than P6. An important factor may explain the difference: the larger the grain size of a crystalline material, the weaker it is. The grain boundary dislocation of small grain size increases the resisting force of plastic flowing. For example, the yield stress follows the Hall-Petch relation (see Equ. 2-10) for a number of steels.

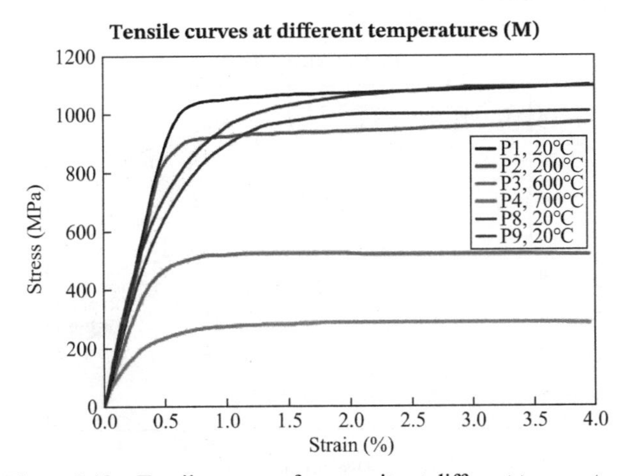

Figure 5.12 Tensile curves of martensite at different temperatures

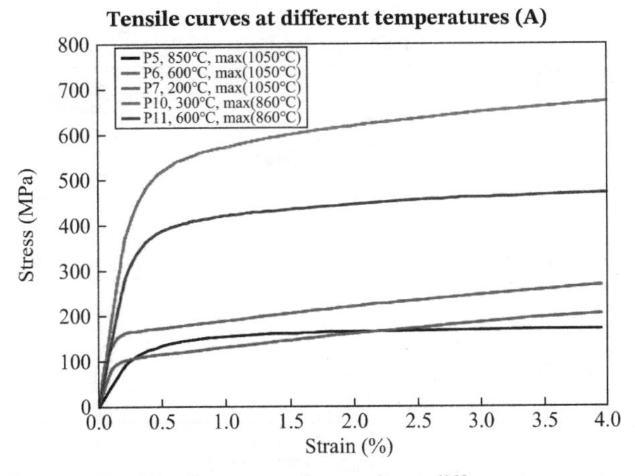

Figure 5.13 Tensile curves of austenite at different temperatures

Eleven figures with load-unload curves were acquired from the corresponding tests (from P1 to P11). Herein, only two examples of cyclical load-unload tests (P1 and P5) are reported here (Figure 5.14 and Figure 5.15), and others are given in Appendix B. The corresponding strain-stress curves are shown in Figure 5.16 and Figure 5.17. Other tensile curves at different temperatures are given in the Appendix B. The test of 15-5PH at room temperature (martensite state) was performed and generated the force-displacement curve shown in Figure 5.14. It shows that the material deteriorated and force decreased with the increasing of displacement after the force reached ultimate strength. At last, material was broken sharply. In terms of 15-5PH at 850°C (austenite state, see as Figure 5.15), it is much weaker than that in room temperature, and the behavior of full failure is less of sudden because material is more ductile. The cyclical load-unload leads to variation of Young's modulus, then cause the measurement of damage variable, the detailed analyses will be given in Section 5.7.4.

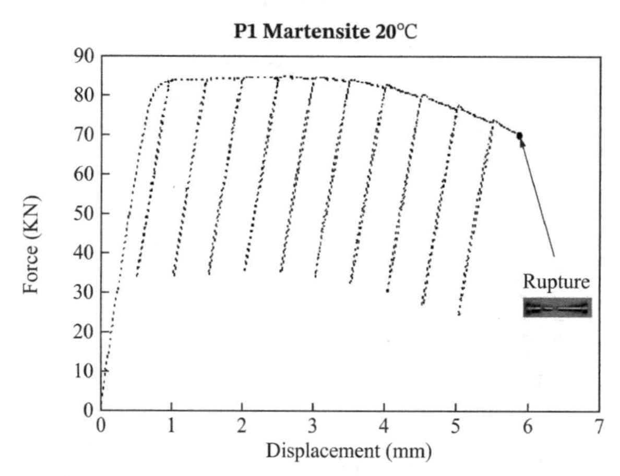

Figure 5.14 Force vs. displacement curves of cyclical load-unload tests at 20°C, martensite state (P1)

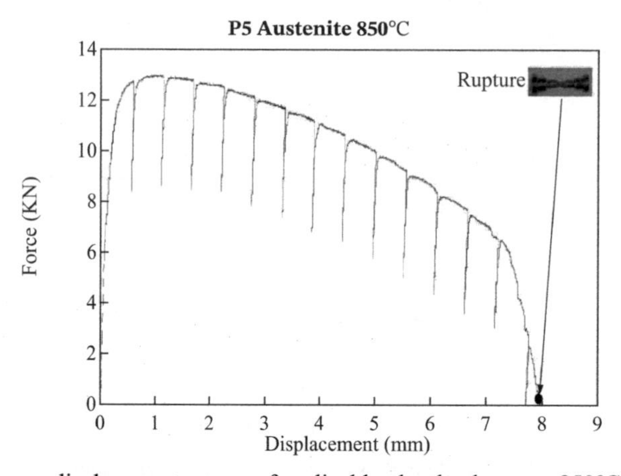

Figure 5.15 Force vs. displacement curves of cyclical load-unload tests at 850°C, austenite state (P5)

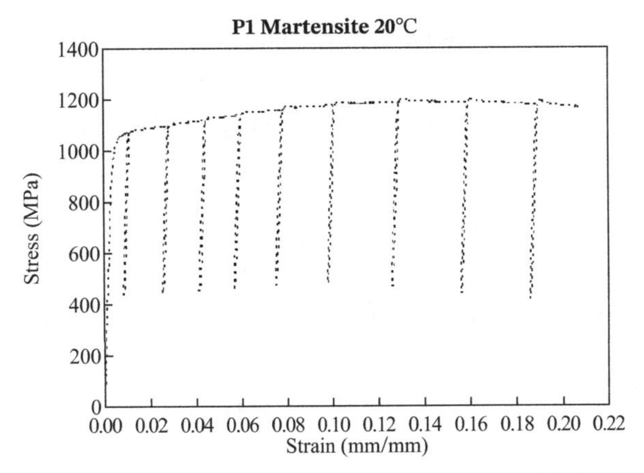

Figure 5.16 Stress vs. strain curves of cyclical load-unload tests at 20°C, martensite state (P1)

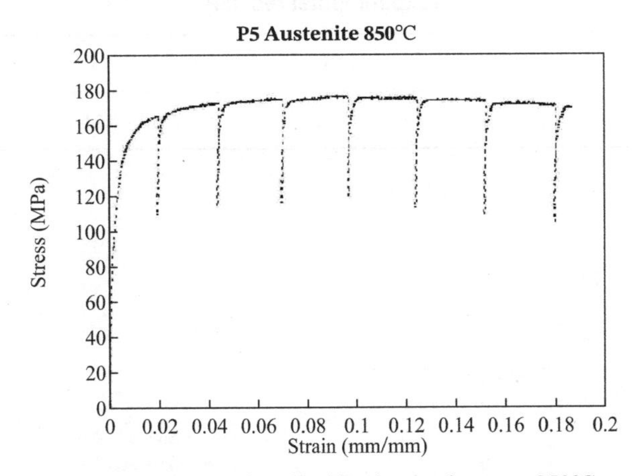

Figure 5.17 Stress vs. strain curves of cyclical load-unload tests at 850°C, austenite state (P5)

5.7.3 Identification of parameters of transformation plasticity model

In terms of phase-transformation model (Magee model), there are 2 material parameters included: M_s and β. M_s can measure directly from the curve of free expansion, and parameter β should do fitting process to experimental results. The results are given in Table 5.4.

As far as transformation induced plasticity model (Leblond model) consideration, it contains 2 parameters: $\Delta\varepsilon_{\gamma-\alpha}^{th}$ and $\sigma_{\gamma}^{y}.\Delta\varepsilon_{\gamma-\alpha}^{th}$ can be measured directly from the curve of free expansion, and its result is listed in Table 10. The parameter σ_{γ}^{y} should be identified from experimental data.

The Lelond's transformation plasticity model (see Section 2.6.2) provides increment of transformation induced plasticity. In order to identify the parameter σ_{γ}^{y}, the integral form is given as:

$$\varepsilon^{pt} = -\frac{3\Delta\varepsilon_{\gamma-\alpha}^{th}}{\sigma_{\gamma}^{y}} \cdot S \int_{0}^{1} \ln z \cdot \mathrm{d}z \tag{5-7}$$

where, $S = \frac{2}{3}\sigma_{eq}$ in one dimensional tensile case.

$$\varepsilon^{pt}(z) - \varepsilon^{pt}(0) = -\frac{2\Delta\varepsilon_{\gamma-\alpha}^{th}}{\sigma_{\gamma}^{y}} \cdot \sigma_{eq}[z(\ln(z) - 1)]_0^z \tag{5-8}$$

$$\varepsilon^{pt}(1) - \varepsilon^{pt}(0) = \frac{2\Delta\varepsilon_{\gamma-\alpha}^{th}}{\sigma_{\gamma}^{y}} \cdot \sigma_{eq} \tag{5-9}$$

Through fitting the data in the Table 5.5, the parameter σ_{γ}^{y} was identified to be: (see Figure 5.19).

$$\sigma_{\gamma}^{y} = 217 \pm 0.5\text{MPa} \tag{5-10}$$

Phase-transformation induced plasticity under stress loads (experimental results) Table 5.5

P (MPa)	37.5	50	75	115.5
ε^{pt} (%)	0.22	0.27	0.54	1.01

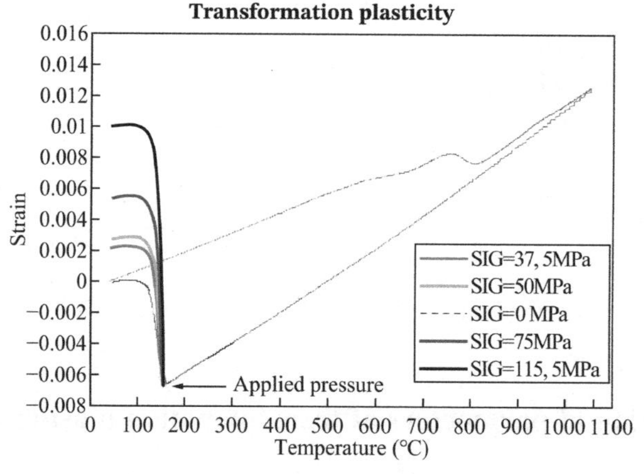

Figure 5.18 TRIP from the experiment with heating and cooling loading (only thermal and plastic stain herein, elasticity is moved)

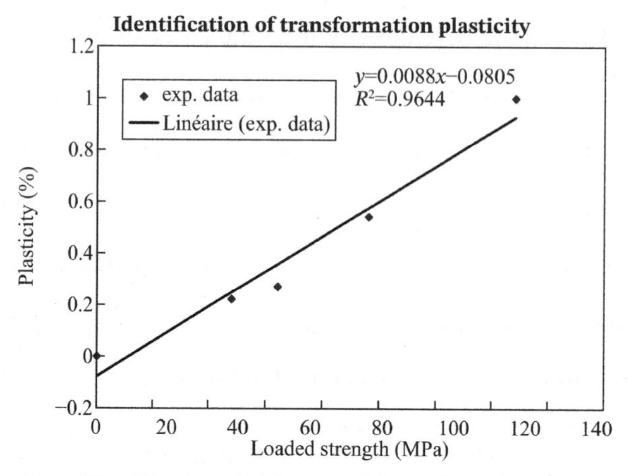

Figure 5.19 Identification of parameters of transformation plasticity model

5.7.4 Identification of parameters of damage model

The damage model (Lemaitre model and the proposed model in Chapter 4) requires several parameters to be identified [124][18]. These constants and their explanation are as follows:

D_0: initial damage. The initial damage is difficult to determine because it is strictly related to the inclusions distribution in the virgin material microstructure. A scanning electron microscope (SEM) investigation can provide an idea of the initial particle distribution and amount of damage in the strained area. As usual, this parameter could be assumed to be zero for a virgin material or at the beginning stage of the damage calculation. For example, we can measure the damage area in Figure 5.32a. A rough value of initial damage is estimated to be: $D_0 \approx 0.001$.

p_D: threshold damage plastic strain. Plasticity damage starts to occur only after a specific strain level. Below this strain level the material microstructure behaves as a continuum. Microvoids can nucleate either by debonding of the included particles from the ductile matrix or by particle breaking.

p_R, D_c: rupture plastic strain and critical damage. When this critical strain is reached, failure occurs. p_R is directly related to the critical amount of damage that critically reduces the load carrying capability of the effective resisting section. In theory, when failure occurs the critical damage variable D_c should be equal to 1. In fact, Experimental observations of almost metals show that failure occurs before $D = 1$.

κ: damage exponent. This constant carries information guiding the type of damage evolution and can be determined from tensile tests. The value gives the degree of nonlinearity of the damage evolution law. For a given p_D, p_R, D_0 and D_c, a discriminating κ value exists which determines the convexity of the damage evolution as a function of plastic strain.

If D_0 is assumed to be equal to 0, Equ.4-25 writes:

$$D = \frac{D_c}{(p_R - p_D)^\kappa}\langle p - p_D\rangle^\kappa \tag{5-11}$$

The exponent κ is determined as the slope of the best fitting line of the experimental damage measurements given by:

$$\ln(D) = \ln(D_c) + \kappa \cdot \ln\left\langle\frac{p - p_D}{p_R - p_D}\right\rangle \tag{5-12}$$

And then we have:

$$\ln(D) = \kappa \cdot \ln(p - p_D) - C \tag{5-13}$$

with

$$C = \kappa \cdot \ln(p_R - p_D) - \ln(D_c) \tag{5-14}$$

where the constant C is the intersection of the fitting line with the ordinate axis. Usually the failure strain and critical damage can be measured pretty well, but the threshold strain p_D is not always so easy to determine as a result of the experimental scatter.

Once the slope is found, it is possible to have a good estimation of the form of previous expression as

$$D_c = \exp[\kappa \cdot \ln(p_R - p_D) - C] \tag{5-15}$$

Through fitting the experimental results, it is not difficult to identify the parameters of damage model (see Figure 5.20). Since the parameters are calibrated (Table 5.6), the evolutions of damage provided by constitutive equations are obtained. For example, Figure 5.20 displays the comparison between the fitting curve and experimental results of damage evolution in martensite state at 20°C. Figure 5.20 shows the material is more difficult to be damaged (damage evolution curves lower) while temperature increases. Comparing damage evolution curves of P 1 (maximum temperature: 20°C), P8 (maximum temperature: 1050°C), P9 (maximum temperature: 860°C) at room temperature, higher maximum temperature experienced by material make it to have better ductility. It may be explained by the change of material characteristic: there remain small quantities of austenite after martensitic transformation in P8 and P9, the higher maximum temperature leads grain size coarsening of austenite. Damage evolution laws strongly depend on temperature history if the transformation temperature is exceeded. Figure 5.21 gives the damage fitting of P1 and the fitting of other specimen can be found in Appendix B.

The damage parameters identified by experiment of round bar are limited in the martensite state. In the tests of austenite, the decrease of Young's modulus is not observed within region of 20% strain. The measurements of damage over 600°C by cyclical load-unload tests becomes very difficult because the very high temperature enhance the ductility of the material and also because the occurrence of significative viscous behaviour.

Identified damage parameters of 15-5PH (martensitic phase) Table 5.6

Temperature	p_D	p_R	D_c	C	κ
P1, 20°C	0.010	0.21	0.18	1.02	0.42
P2, 200°C	0.012	0.20	0.08	1.23	0.79
P3, 600°C	0.020	0.19	0.06	0.97	1.05
P8, 20°C	0.012	0.24	0.16	1.30	0.66
P9, 20°C	0.012	0.23	0.17	1.03	0.5

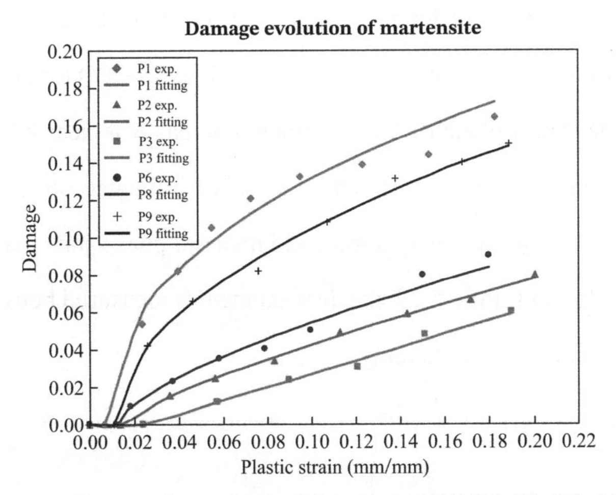

Figure 5.20 Fitting of damage evolution at martensitic state, at 20°C (P1, P8, P9), 200°C (P2), 600°C (P3)

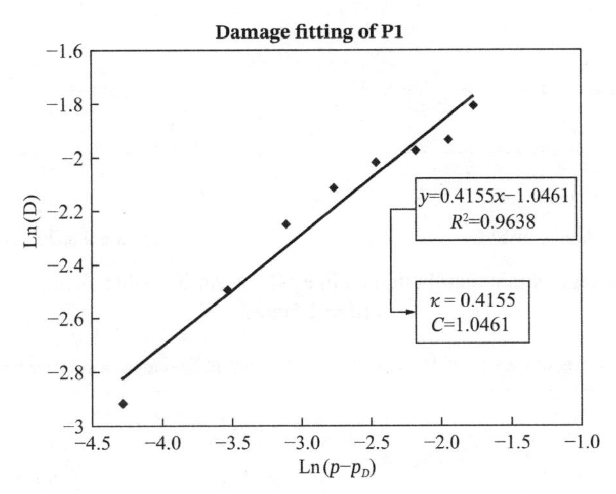

Figure 5.21 Identification of damage exponent of damage model (P1)

5.7.5 Comparative analysis of flat notched specimen

The evolutions of strain localization on surface with speckle pattern of flat notched specimen (FNS) are obtained by digital image correlation of tensile experiments of FNS. Three cases (Case

A, B, C) with different notches and five thermal conditions are analyzed in order to compare study the strain on surface of FNS.

Figure 5.26a and Figure 5.23a show that the notch radius has a large influence onto the axial strain (EPYY) along line MN (mid section, see Figure 5.7a). However, it has less influence on transversal strain (EPXX) (Figure 5.26a). Largest strain always lies in edge of notch what ever its radius. Figure 5.26 b and Figure 5.23 b indicate that material state influences distribution of strain on surface of specimen because different phases have different ductility. Temperature does not seem to have an important effect on strain distribution. Along line PQ (axial line, see Figure 5.7a), largest strain occurs near mid plane of the specimen whereas it is near edge of notch along line MN. Figure 5.28a shows that radius of notch has a small influence on the overall tensile curve (displacement vs. force). However, temperature and material phase state play an important role in tensile test (Figure 5.28b). In Figure 5.29, the displacement is measured between point M and point N and it is increasing with the force loading.

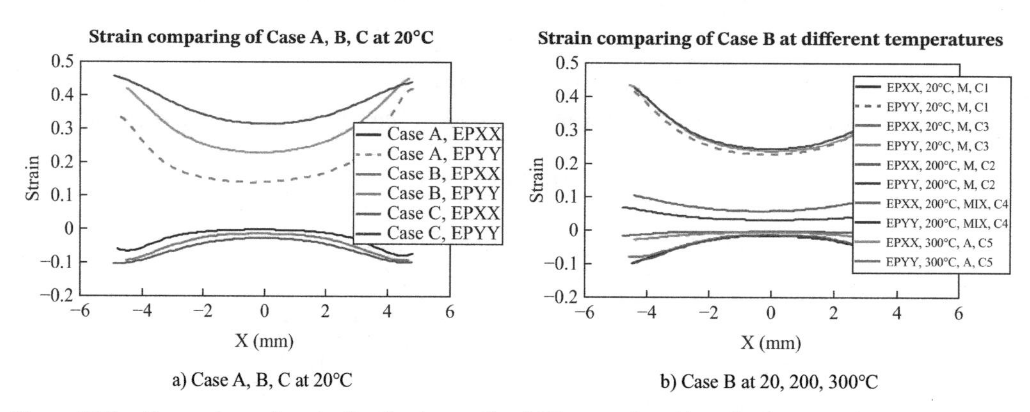

a) Case A, B, C at 20°C

b) Case B at 20, 200, 300°C

Figure 5.22 Comparison of strain distribution on line MN on surface when displacement is equal to 1.5mm (d = 1.5mm)

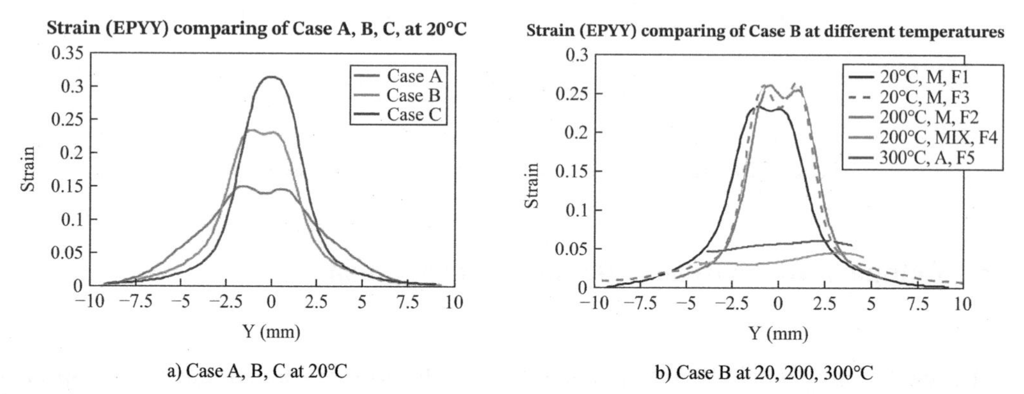

a) Case A, B, C at 20°C

b) Case B at 20, 200, 300°C

Figure 5.23 Comparing strain distribution (EPYY) on line PQ on surface when displacement is equal to 1.5mm (d = 1.5mm)

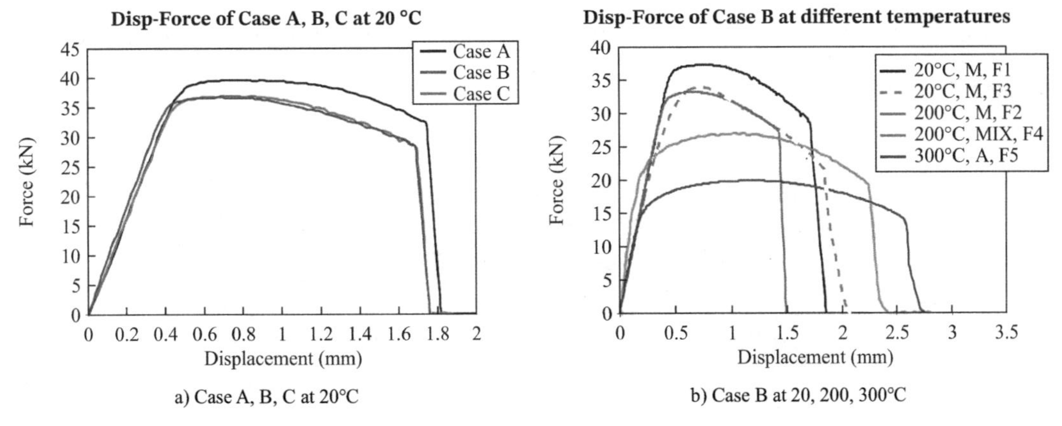

a) Case A, B, C at 20°C

b) Case B at 20, 200, 300°C

Figure 5.24 Displacement vs. force of tensile tests of FNS

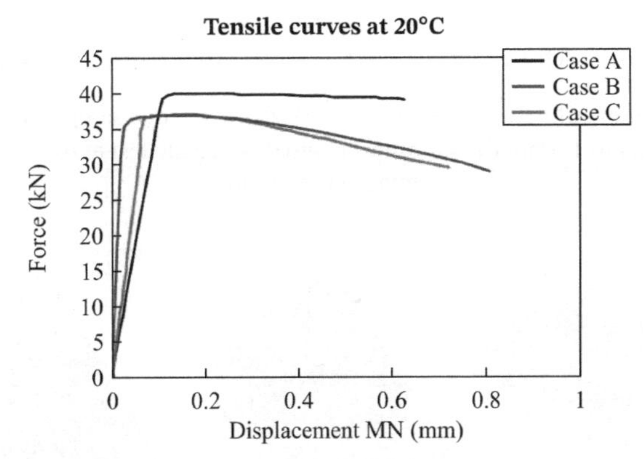

Figure 5.25 Displacement (between point M and point N, mid section) vs. force of tensile tests of FNS at 20°C

Figure 5.26 shows that thickness reductions are quite different in various flat specimens depending on the notch radii: the maximum thickness reduction is near the notch edges for Case A, somewhere inside for case B and at the center for case C. Figure 5.27 shows that the principal strains (in longitudinal direction, EPYY) of three cases have the similar distribution along line MN: peak value of strain always occurs at the notch edges. The reduction of thickness could be used as rough damage evolution estimation, and we could consider that maximum damage lies in the place where minimum section appears. Therefore, the conclusion resulting from the comparison of Figure 5.26 and Figure 5.27 is that the largest strain is not at the same location as maximum thickness reductions. The strains are rather large near failure (> 10%). The measured strains are mainly elastoplastic because of incompressibility. It is an indication that damage is developing. The elastoplastic computations of Case A to Case C show that the peak values of triaxial stress ratio move from edges to center, thus is in agreement with the observations mentioned above. The experimental results leaded to the conclusion that the stress triaxiality plays a dominating role in the evolution of damage.

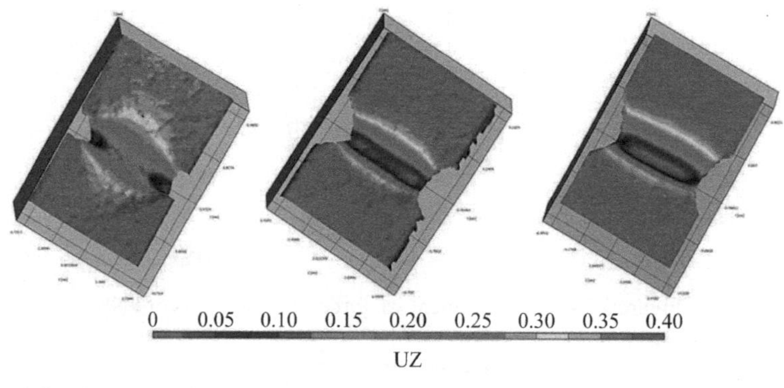

a) Case A R = 1mm (F1A) b) Case B R = 2.5mm (F1B) c) Case C R = 4mm (F1C)

Figure 5.26 3-D distribution of vertical displacement (UZ) on notched samples (F1A, F1B, and F1C)when loading displacement is equal to 1.5mm (d = 1.5mm) in uniaxial tensile test at room temperature by using digital image correlation

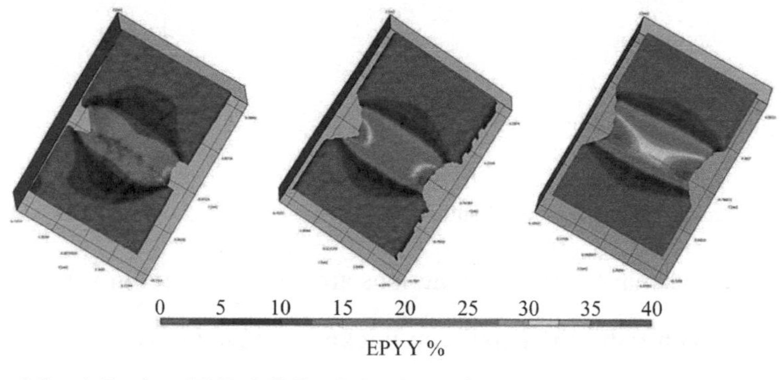

a) Case A R = 1mm (F1A) b) Case B R = 2.5mm (F1B) c) Case C R = 4mm (F1C)

Figure 5.27 Distribution of longitudinal strain (EPYY) on notched samples (F1A, F1B, and F1C) when loading displacement is equal to 1.5mm (d = 1.5mm) in uniaxial tensile test at room temperature by digital image correlation

Figure 5.28 and Figure 5.29 show that the displacement and strain are similarly distributed on surface of FNS at different temperatures because the notch geometry decides strain gradients from the specimen center to the notch edge. However, there is a quantitative difference: displacement Z or strain EPYY at martensitic phase state (Figure 5.28a, b and Figure 5.29a, b) have larger level than their in austenitic phase (Figure 5.28c and Figure 5.29c) or at mixture state (Figure 5.28d and Figure 5.29d). In other words, severe necking can be observed in martensitic state, but not in austenitic state when the displacement is loaded up to 1.5 millimeters.

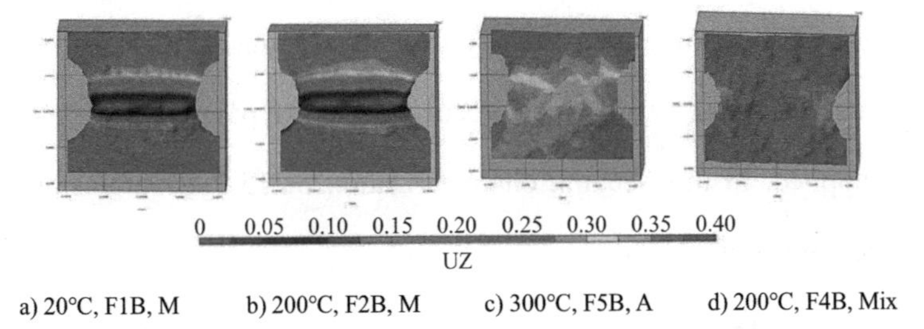

a) 20°C, F1B, M b) 200°C, F2B, M c) 300°C, F5B, A d) 200°C, F4B, Mix

Figure 5.28 3-D distribution of displacement of Case B with different temperature histories (F1B, F2B, F5B and F4B) when loading displacement is equal to 1.5mm (d = 1.5mm) by using digital image correlation. (A is austenite; M means martensite; Mix indicates in mixed phase state.)

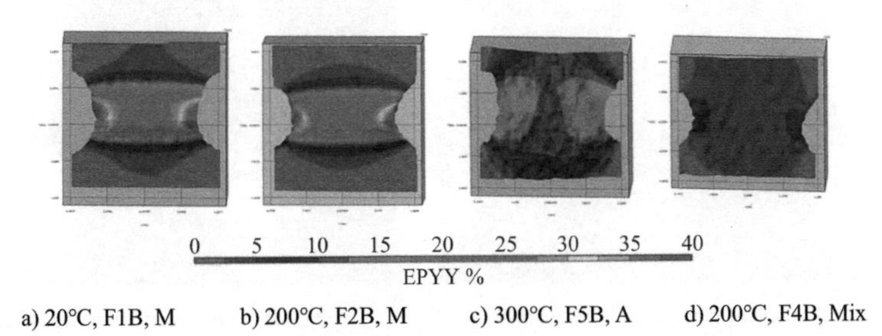

a) 20°C, F1B, M b) 200°C, F2B, M c) 300°C, F5B, A d) 200°C, F4B, Mix

Figure 5.29 Distribution of strain in longitude direction (EPYY) of Case B with different temperature histories (F1B, F2B, F5B and F4B) when loading displacement is equal to 1.5mm (d = 1.5mm) by using digital image correlation

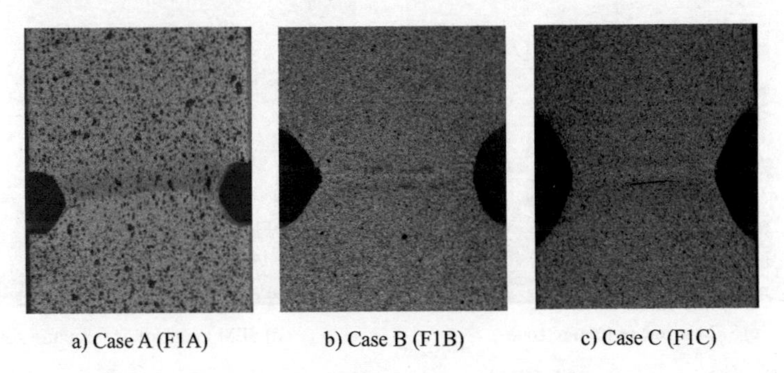

a) Case A (F1A) b) Case B (F1B) c) Case C (F1C)

Figure 5.30 Macroscopic fracture observation of FNS with various notches at room temperature at the moment of rupture (Case A, Case B and Case C)

The images at moment of rupture are traced by digital camera (Figure 5.30). Figure 5.30c shows the initial crack occurs at center and then extends to both edges for Case C. This is in good agreement with the distribution of maximum damage level of Case C. For Case A and Case B, the initial crack could not be captured because the specimens break into two parts too quickly.

5.8 Microstructure characterization of 15-5PH

Microscopic observation is performed due to study damage mechanism at micro scale and to provide an indication for the choice of an appropriate damage models. Microstructure characterization can be quite different due to the various methods of mechanical process as well as the differences of thermal histories applied. In this section, characterization of 15-PH martensitic stainless at room temperature will be discussed in detail.

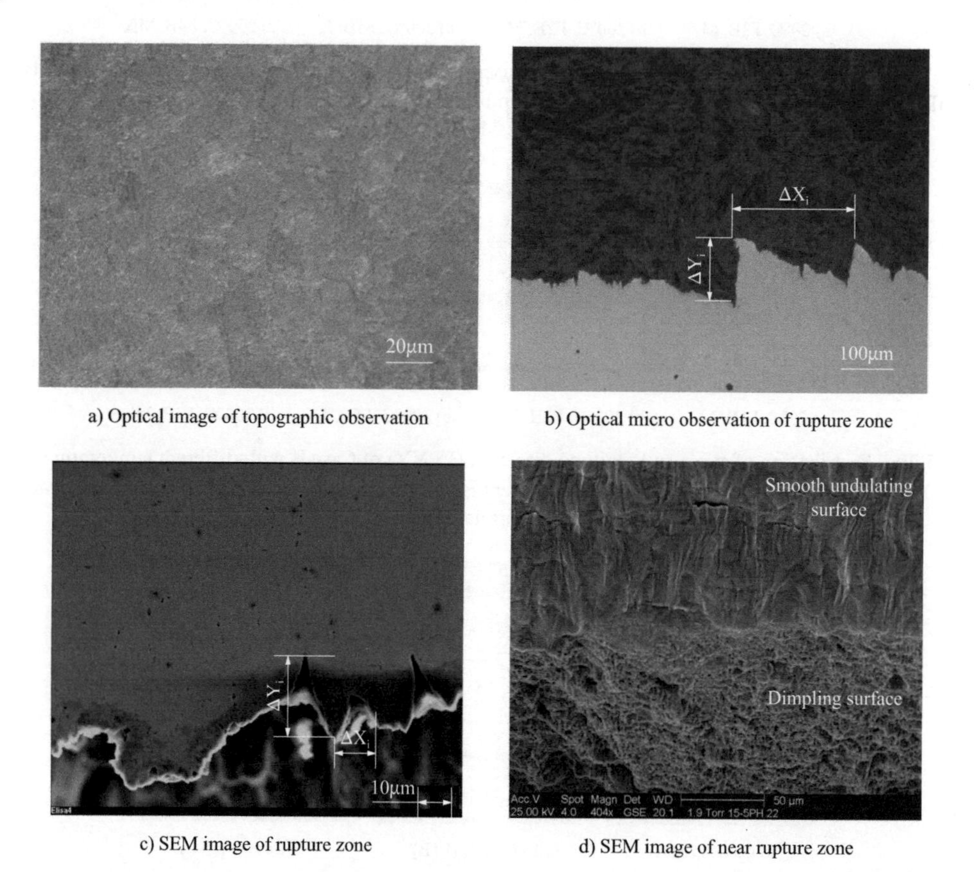

a) Optical image of topographic observation

b) Optical micro observation of rupture zone

c) SEM image of rupture zone

d) SEM image of near rupture zone

Figure 5.31 Micro observation of 15-5PH by optical microscopy and scanning electron microscopy (SEM)

Figure 5.31a gives a topographic observation of 15-5PH before tensile tests are performed. It shows that the microstructure is in form of lathing martensite at room temperature and diameter of grain sizes are between 10 and 20 micron. Figure 5.31b and Figure 5.31c show the fracture surface (longitudinal cross-section) of RB after tensile test, and the rupture zone with delamination are respectively observed by using optical microscopy and SEM. Figure 5.31 b show the edge of rupture presents sawtooth-like shape. The fracture is torn not at conventional 45° but at axial direction, and then extended to 0.15 mm deep. A typical broken section with characteristic dimensions is presented in Figure 5.31b, ΔX_i is the length of horizontal segment, perpendicular to the strength direction. The length of vertical segment parallel to the strength loading is denoted by ΔY_i. They were measured from the Figure 5.31b: $\Delta X_i = 260\mu\text{m}$, $\Delta Y_i = 130\mu\text{m}$ and $\Delta X_i/\Delta Y_i = 2$. In order to construct a model of mesomechanism, it is important to know the distribution of ΔX_i and ΔY_i all along the entire cross-section of a target, and provides element dimensions of mesh of mesoscopic model. Δx_i and Δy_i in Figure 5.31c can be a reference for dimensions of microscopic model. For example, herein has $\Delta x_i = 12\,\mu\text{m}$, $\Delta y_i = 22\,\mu\text{m}$ and $\Delta x_i/\Delta y_i = 0.55$. Compared the data of Figure 5.31b, it is found that the effect of dimension is obvious: $\Delta X_i/\Delta Y_i > \Delta x_i/\Delta y_i$, i.e. the fractography is different between at mesoscale and at microscale. In fact, it is necessary to do statistical analysis of all segments along the broken section in order to get an average value.

Voids or cavities with diameter of 1-2 microns can be observed near the edge of rupture in Figure 5.31c. The high magnification view of the fracture surface in Figure 5.31d clearly shows the smooth delamination surface as well as the dimpled ductile fracture area. The smooth undulating surface suggests some type of decohesion of the grain boundaries. Micro cracks can be observed in the smooth delamination zone. Microvoids have higher density and larger dimensions near the broken section according to higher plastic strain.

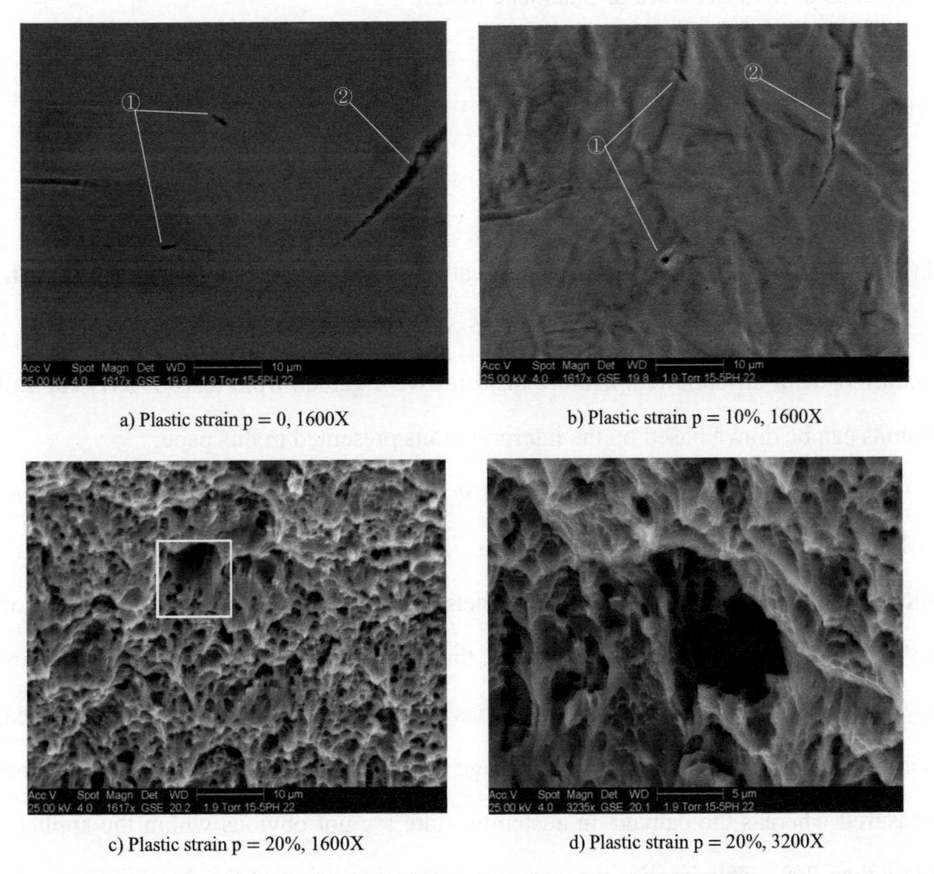

a) Plastic strain p = 0, 1600X

b) Plastic strain p = 10%, 1600X

c) Plastic strain p = 20%, 1600X

d) Plastic strain p = 20%, 3200X

Figure 5.32 Scanning electron microscope (SEM) images of 15-5PH specimens when various strains are loaded by using SEM at room temperature (thermal condition C1). Figure d) is a zoom of the square area marked in Figure c). The circle 1 shows microvoids observed and the circle 2 shows a microcrack

Figure 5.32 shows microstructures of 15-5PH under various deformations. Before deformation is performed (Figure 5.32a), the material microstructure is quite smooth except a few microvoids and microcracks that can be regarded as initial damage. With the loading of deformation, grains are twisted and surface of microstructure are in concavo-convex shape to some extent (Figure 5.32b). With higher deformation, dimples can be observed in the surface in Figure 5.32 and dimpled ductile damage happen. Figure 5.32d shows a larger void with coalescence because of the application of higher magnification.

The experimental observation of microstructure provides basic information of damage mechanism under tensile condition. This investigation helps to choose or develop a damage model in order to simulate damage and residual stress. The typical microvoids and microcracks, which were observed in microscopic experiment, are basic concept of damage Kachanov in RVE. The observed distribution and evolution of microvoids verify the ductile damage induced by plastic strain, which is used in Lemaitre & Chaboche model.

5.9 Conclusion

In this Chapter, we discussed the identification of the parameters of damage model, macro cracks of tensile tests and microstructure of 15-5PH. In addition, mechanical properties at high temperature or with different temperature histories were gained from tests of RB. The following conclusions can be drawn based on the interpretations presented in this paper:

(1) Damage variables can be measured by variation of Young's modulus resulting from circularly load-unloading tests of round bar. The careful processing of data leads to the identification of the parameters of damage models, which we have chosen (Lemaitre's model and Borona's model). The processed results indicate that 15-5PH in martensitic state at low temperature are easy to damage. When the same specimen has been heated a high temperature the damage rate is less important. In a series of tests, the damage in martensitic state of 15-5PH can be observed and measured whereas the damage in austenitic state are not obvious within the applied plastic strain less than 20%. This implies the austenite have large value of the threshold strain to cause damage. However, this does not mean the threshold strain is over 20%. A SEM investigation should

fill in the absence of damage parameters of austenite phase. It is reasonable to disregard the damage in austenite phase in simulation of multiphase in limited deformation condition, typical of welding situation.

(2) Expansion-contraction tests with applied stress allowed successful identification of phase transformation (Koistinen and Marburger) and fitting of TRIP (Leblond model).

(3) Damage and rupture in macroscopic scale are observed in the tensile tests of FNS. The strain localization is analyzed using digital image correlation. The results show that largest damage does not always occur in the place of the largest strain because damage is a competitive result between plasticity and triaxial stress ratio.

(4) Comparison of tests with various temperature histories shows that the mechanical behaviors (e.g. strain-stress curve) including damage are affected by histories of thermal loading. The damages are strongly affected by high temperature. Therefore, the effect of temperature history should be taken into account in modeling.

(5) Grain size, rupture zone, microvoids and microcracks are characterized in the microscopic measurement. Microvoids grow with increasing of deformation. Dimples are observed to be the dominant features and were formed as a result of the initiation and coalescence of microvoids. Thus provides a justification of the choice of Lemaitre and Borona damage models associated to microvoids and microcracks based on continuum damage mechanics. The information is also necessary to understand better the damage behaviors at macroscale.

Since the parameters of phase transformation, transformation plasticity, and damage models are successfully calibrated through experimental methods, the numerical simulation can be implemented in next chapter.

Chapter 6 Numerical Simulation and Implementation of Constitutive Equations

6.1 Introduction

In this chapter, we introduce numerical analysis of structures under thermo mechanical loading with damage and phase transformation. On the one hand, the numerical calculation of notched specimen is done and compared with the experimental results. On the other hand, an example of disk welding is given in order to illustrate the particularity of numerical simulation of welding through the two-scale model, which is proposed in the previous sections.

We focus our study on the metallurgy and damage in mechanical calculation. It is supposed that the temperature field is already calculated, and proportions of phase are calculated from temperature field, using standard methods of these metallurgical computations (Koistinen and Marburger's empirical law, Leblond's model...).

In addition, in order to apply to the welding numerical prediction, the constitutive equations have to be implemented in a finite element program. The proposed model is implemented in computing software CEA CAST3M® finite element code, which integrates not only the computing processes themselves but also the functions of construction of the model (pre-processor) and the functions of processing of the results (post processing). The CAST3M is a proper numerical platform, which provides a large choice of mechanical models or laws.

6.2 Metallurgical calculation

6.2.1 Phase transformation calculation

The calculation of austenite phase transformation (austenitization) can use the phenomenological model (Equ.2-4) proposed by Leblond, which was introduced in Section 2.3.4. In the case of welding simulation of damage and stresses, we can use its simple form at equilibrium state. For a slow heating rate, there is enough time to make the austenite fraction to reach equilibrium state for each temperature, and the proportion of austenite is approximately linear with the temperature during phase transformation (Figure 2.8).

For martensitic phase transformation, Koistinen and Marburger's empirical law (Equ.2-3) can be applied to calculate the proportion of martensite. In the law, proportion of martensite is a function of temperature and austenite's proportion.

$$z_1 = z_2\left(1 - e^{-\beta<M_s-T>}\right) \tag{6-1}$$

where z_1 is volume fraction of martensite and z_2 represents volume fraction of austenite; M_s is martensite start temperature; β is coefficient depend on material; T represents temperature.

In martensitic transformation, on one hand, there is no differential equation to integrate. On the other hand, it is necessary to take into account the irreversibility of the transformation, i.e. martensite can not be reversed when temperature is below Ac_1. Practically, the Equ. 6-1 can be written as:

$$z_1(t+\Delta t) = \max\left(z_1(t+\Delta t), z_1(t)\right) \tag{6-2}$$

with

$$z_{1'}(t+\Delta t) = z_2(t+\Delta t)[1 - \exp(-\beta\langle M_s - T(t+\Delta t)\rangle)] \tag{6-3}$$

$$z_2(t+\Delta t) = 1 - z_1(t) \tag{6-4}$$

In the martensitic phase computation, the input variables include: $z_1(t)$, $z_2(t)$ and $T(t+\Delta t)$. The calculation of metallurgical transformation is performed at Gauss point of FEM mesh.

Therefore, the metallurgical calculation can be summarized in the following:

Phase transformation calculation

If $(\dot{T} \geqslant 0$ and $T \leqslant Ac_1)$ or $(\dot{T} \geqslant 0$ and $T \geqslant Ac_3)$

No phase transformation

If $\dot{T} \geqslant 0$ and $Ac_3 > T > Ac_1$

Austenitization

If $(\dot{T} < 0$ and $T > M_s)$ or $(\dot{T} < 0$ and $T < M_f)$

No phase transformation

If $\dot{T} < 0$ and $T > M_s$

Martensitic transformation

In addition, the parameters of phase transformation of 15-5PH are given in the Chapter 5.

6.2.2 Grain size calculation

The growth of grain size can be calculated through conventional equation proposed by Alberry, Ikawa, and Leblond (see Section. 2.4.1). The following flow chart summarises the method:

Grain size calculation

If $\dot{z}_2 > 0$

$$\frac{d}{dt}(G^a) = C \cdot \exp\left(-\frac{Q}{RT}\right) - \frac{\dot{z}_2}{z_2} G^a \tag{6-5}$$

$$\text{or } \frac{d}{dt}(z_2 G^a) = z_2 \cdot C \cdot \exp\left(-\frac{Q}{RT}\right) \tag{6-6}$$

If $\dot{z}_2 \leqslant 0$

$$\frac{d}{dt}(G^\alpha) = C \cdot \exp\left(-\frac{Q}{RT}\right) \tag{6-7}$$

where

G : grain size

T : absolute temperature

R : constant for gases

a, C, Q are positive constants. (Generally, $a = 4$, $C = 0.4948 \times 10^{14} \text{mm}^4/\text{sec}$, $Q/R = 63900$ gives satisfactory results.)

The Equ. 6-5 can be written as following form:

$$\frac{d}{dt}(G) = f[G(t), V_i(t)] \tag{6-8}$$

where $V_i(t)$ represents other variables except G.

The implicit differential equation is integrated by Runge-Kutta method. The austenite grain size at time $t + dt$ is calculated by:

$$G(t + dt) = G(t) + \frac{dt}{6}\left[\dot{G}_1 + 2\left(\dot{G}_2 + \dot{G}_3\right) + \dot{G}_4\right] \tag{6-9}$$

with

$$\dot{G}_1 = f[G(t), V_i(t)] \tag{6-10}$$

$$\dot{G}_2 = f\left[G(t) + \dot{G}_1 \frac{\mathrm{d}t}{2}, V_i\left(t + \frac{\mathrm{d}t}{2}\right)\right] \tag{6-11}$$

$$\dot{G}_3 = f\left[G(t) + \dot{G}_2 \frac{\mathrm{d}t}{2}, V_i\left(t + \frac{\mathrm{d}t}{2}\right)\right] \tag{6-12}$$

$$\dot{G}_4 = f\left[G(t) + \dot{G}_3\,\mathrm{d}t, V_i(t + \mathrm{d}t)\right] \tag{6-13}$$

6.2.3 Calculation of transformation plasticity

In the case of numerical simulation of residual stresses and damages induced by welding, only martensitic transformation plasticity is considered because austenite transformation occurs at high temperature when the material has more ductility and less stress concentration. The martensitic transformation (1→2) plasticity can be calculated by Leblond's model presented in Section 2.6.2. The simple expression is written as:

$$\dot{E}^{pt} = -\frac{3\Delta\varepsilon_{1-2}^{T_{ref}}}{\sigma_2^y} \cdot S \cdot h\left(\frac{\Sigma^{eq}}{\Sigma^y}\right) \cdot (\ln z_2) \cdot \dot{z}_2 \forall z_2 \subset [0.03,\quad 1] \tag{6-14}$$

Its incremental form is:

$$\Delta E^{pt}(t) = -\frac{3\Delta\varepsilon_{1-2}^{T_{ref}}}{\sigma_2^y} \cdot S(t) \cdot h\left(\frac{\Sigma^{eq}}{\Sigma^y}\right) \cdot [\ln z_2(t)] \cdot [z_2(t+\Delta t) - z_2(t)] \tag{6-15}$$

$$E^{pt}(t+\Delta t) = E^{pt}(t) + \Delta E^{pt}(t) \tag{6-16}$$

The experiments presented in the Chapter 5 have given the temperature-independent parameters: $\Delta\varepsilon_{1-2}^{T_{ref}}$, σ_2^y. The parameter $h(\Sigma^{eq}/\Sigma^y)$, which translates the non-linearity of the transformation plasticity according to the equivalent macroscopic stress applied, has been introduced in Section 2.6.2 and Section 2.7.2 in detail.

6.3 Numerical implementation of the two-scale model

6.3.1 Introduction

As far as the numerical analysis is concerned, in our simulation of the two-scale model proposed in Chapter 4, the first step is to compute the coupled transient temperature and metallurgical phase field. The second one is stress, strain and damage field computation and then the states of each phase (stresses, strains, damages, internal variables …) are output. At the end, we can homogenize the strains and stresses, displacements and damages. The calculation of temperature field is not discussed in our study, and the calculation metallurgical phase field was given in the previous section. Here, we focus our study on mechanical analysis in two-scale model.

6.3.2 Algorithm

The algorithm in small displacements is given in the following part (Figure 6.1). The model presented previously relates only to the integration of the constitutive law of multiphase material.

If the incremental theory is applied to solve the problems of multiphase behaviors, the increment of macroscopic stress $\Delta\Sigma_p$ between step p and step $p+1$ can be calculated from the following equations. Here, 1 and 2 represent respectively martensite and austenite as the previous convention.

$$
\begin{aligned}
\Delta\Sigma_{(p)} &= \Sigma_{(p+1)} - \Sigma_{(p)} = \sum_{i=1,2} \mathrm{z}_{i(p+1)}\sigma_{i(p+1)} - \sum_{i=1,2} \mathrm{z}_{i(p)}\sigma_{i(p)} \\
&= \sum_{i=1,2} \mathrm{z}_{i(p+1)} H_{(p+1)} \varepsilon^{e}_{i(p+1)} - \sum_{i=1,2} \mathrm{z}_{i(p)} H_{(p)} \varepsilon^{e}_{i(p)} \\
&= \sum_{i=1,2} \mathrm{z}_{i(p+1)} H_{(p+1)} \left[\varepsilon^{e}_{i(p+1)} - \varepsilon^{e}_{i(p)}\right] + \sum_{i=1,2} \left[\mathrm{z}_{i(p+1)} H_{(p+1)} - \mathrm{z}_{i(p)} H_{(p)}\right] \varepsilon^{e}_{i(p)} \\
&= \sum_{i=1,2} \mathrm{z}_{i(p+1)} H_{(p+1)} \left[\Delta\varepsilon^{tot}_{i(p)} - \Delta\varepsilon^{th}_{i(p)} - \Delta\varepsilon^{p}_{i(p)}\right] + \sum_{i=1,2} \left[\mathrm{z}_{i(p+1)} H_{(p+1)} - \mathrm{z}_{i(p)} H_{(p)}\right] \varepsilon^{e}_{i(p)} \\
&= \left[\sum_{i=1,2} \mathrm{z}_{i(p+1)} H_{(p+1)}\right] \Delta E_{(p)} - \sum_{i=1,2} \mathrm{z}_{i(p+1)} H_{(p+1)} \Delta\varepsilon^{th}_{i(p)} - \sum_{i=1,2} \mathrm{z}_{i(p+1)} H_{(p+1)} \Delta\varepsilon^{p}_{i(p)} + \\
&\quad \sum_{i=1,2} \left[\mathrm{z}_{i(p+1)} H_{(p+1)} - \mathrm{z}_{i(p)} H_{(p)}\right] H^{-1}_{(p+1)} \sigma_{i(p)}
\end{aligned}
$$

The load increment is:

$$
\begin{aligned}
\Delta\underline{F}_{(p)} = {} & B^{T}\left[\sum_{i=1,2} \mathrm{z}_{i(p+1)} H_{(p+1)}\right]\left(\Delta E_{(p)} - \Delta E^{pt}_{p+1}\right) - B^{T}\left[\sum_{i=1,2} \mathrm{z}_{i(p+1)} H_{(p+1)} \Delta\varepsilon^{th}_{i(p)}\right] - \\
& B^{T}\left[\sum_{i=1,2} \mathrm{z}_{i(p+1)} H_{(p+1)} \Delta\varepsilon^{p}_{i(p)}\right] + B^{T}\left[\sum_{i=1,2} \left(\mathrm{z}_{i(p+1)} H_{(p+1)} - \mathrm{z}_{i(p)} H_{(p)}\right) H^{-1}_{(p)} \sigma_{i(p)}\right]
\end{aligned}
$$

where B is defined by: $\underline{U}^{T} B \Sigma = \int_{\Omega} \mathrm{tr}[\Sigma E(\underline{U})] \mathrm{d}\Omega$.

Then, it can be written as:

$$K_{(p+1)}\Delta\underline{U}_{(p)} = \Delta F_{(p)} + \Delta F_{(p)}^{th} - \Delta F_{(p)}^{e} + \Delta F_{(p)}^{p} + \Delta F_{(p)}^{pt}$$

with

$$\Delta F_{(p)}^{th} = B^T\left[\sum_{i=1,2} z_{i(p+1)}H_{(p+1)}\Delta\varepsilon_{i(p)}^{th}\right]$$

$$\Delta F_{(p)}^{e} = B^T\left[\sum_{i=1,2} (z_{i(p+1)}H_{(p+1)} - z_{i(p)}H_{(p)})H_{(p)}^{-1}\sigma_{i(p)}\right]$$

$$= B^T\left[\sum_{i=1,2} H_{(p+1)}(z_{i(p+1)}H_{(p)}^{-1}\sigma_{i(p)} - z_{i(p)}H_{(p+1)}^{-1}\sigma_{i(p)}\right]$$

$$\Delta F_{(p)}^{pt} = B^T\left[\sum_{i=1,2} z_{i(p+1)}H_{(p+1)}\right]\Delta E_{(p)}^{pt}$$

If the multiphase is coupled with damage, stiffness matrix H is a function of damage variable D and can be written as $H(D)$.

In the two scale model, the partition of the strains at the microscopic level is preserved, and initial transformation plasticity is supposed to be null.

Initialization:

$$E_{(0)}^{tp} = 0;\ \varepsilon_{1(0)}^{th} = 0;\ \varepsilon_{2(0)}^{th} = \Delta\varepsilon_{\alpha-\gamma}^{T_{ref}}$$

The increment of thermal strain of each phase i:

$$\Delta\varepsilon_{i(p)}^{th} = \alpha_{i(p+1)}(T)\cdot T_{(p+1)} - \alpha_{i(p)}(T)\cdot T_{(p)}$$

The accumulation of thermal strain of each phase i:

$$\varepsilon_{i(q)}^{th} = \varepsilon_{i(p)}^{th} + \Delta_p\varepsilon_{i(p)}^{th}\ \varepsilon_i^{tot} = E - E^{pt}$$

The accumulation of thermal stress of each phase i:

$$\sigma_{i(p)}^{th} = H_{i(p)}\varepsilon_{i(p)}^{th}$$

The mixture of thermal force:

$$F_{(p)}^{th} = \sum_{i=1,2} z_i B^T\sigma_{i(p)}^{th}$$

The residual force:

$$R_{(p)} = F_{(p)}^{ext} - F_{(p)}^{int} + F_{(p)}^{th}$$

The following algorithm gives the evolution of residual balance between iteration j and iteration $j+1$. If an equilibrium point exists and the steps are not too large, the residual displacement can decrease until they reach a very small number ξ through iteration.

Calculation of the increment and accumulation of displacement:

$$\Delta U^{(j)} = K_{(p)}^{-1} R_{(p)}^{th}$$
$$U_{(p+1)}^{(j)} = U_{(p)}^{(j)} + \Delta U^{(j)}$$

The linking force is:

$$F^{lin(j)} = K_{(p+1)} U_{(p+1)}^{(j)}$$

$$\text{with } K_{(p+1)} = B^T \left[\sum_{i=1,2} z_{i(p+1)} H_{i(p+1)} \right] B \text{ and } U^T B \Sigma = \int_\Omega \text{tr}[\Sigma E(U)] \mathrm{d}\Omega$$

The total strain:

$$E_{(p+1)}^{tot(j)} = B U_{(p+1)}^{(j)}$$

The total strain of each phase excluding transformation plasticity:

$$\varepsilon_{i(p+1)}^{tot(j)} = E_{(p+1)}^{tot(j)} - E_{(p+1)}^{pt}$$

The mechanical strain:

$$\varepsilon_{i(p+l)}^{mec(j)} = \varepsilon_{i(p+1)}^{e(j)} + \varepsilon_{i(p+1)}^{p(j)} = \varepsilon_{i(p+l)}^{tot(j)} - \varepsilon_{i(p+1)}^{th}$$

The increment and accumulation of stress of each phase:

$$\begin{aligned}
\Delta\sigma_{i(p+1)}^{(j)} &= H_{i(p+1)} \varepsilon_{i(p+1)}^{e(j)} - H_{i(p+1)} \varepsilon_{i(p+1)}^{e(j)} \\
&= H_{i(p+1)} \left[\varepsilon_{i(p+1)}^{mec(j)} - \varepsilon_{i(p+1)}^{p(j)} \right] - H_{i(p)} \left[\varepsilon_{i(p)}^{mec(j)} - \varepsilon_{i(p)}^{p(j)} \right] \\
\sigma_{i(p+1)}^{(j)} &= \sigma_{i(p+1)}^{(j)} + \Delta\sigma_{i(p+1)}^{(j)}
\end{aligned}$$

The internal force:

$$F^{int(j)} = B^T \Sigma^{(j)} = B^T \sum_{i=1,2} z_i \sigma_{i(p+1)}^{(j)}$$
$$R^{(j)} = F^{ext(j)} - F^{int(j)} + F^{lin(j)}$$

TEST

$$\left\| \underline{R}^{(j)} \right\| \leqslant \xi$$

IF NOT

$$\Delta\Delta \underline{U}_{(p)}^{(j)} = K_{(p+1)}^{-1} R^{(j)}$$
$$\Delta \underline{U}_{(p)}^{(j+1)} = \Delta \underline{U}_{(p)}^{(j)} + \Delta\Delta \underline{U}_{(p)}^{(j+1)}$$

IF YES

Calculation of the transformation plasticity at next step:

$$\Delta E_{(p)}^{tp} \text{ and } E_{(p+1)}^{tp} = E_{(p)}^{tp} + \Delta E_{(p)}^{tp}$$

Finish equilibrium iteration.

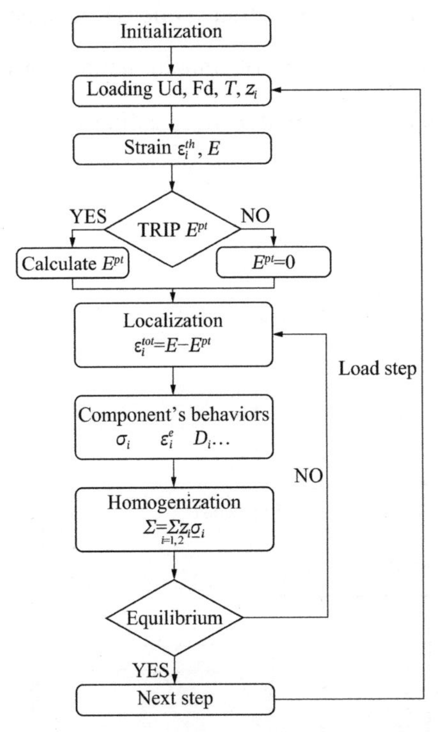

Figure 6.1 Flow chart of multiphase calculation

6.4 Numerical verification of the models

Since the constitutive equations of model are given in the Chapter 4, the numerical calculation are implemented to verify the models and the calculated results are compared to experimental results, which were accomplished in Chapter 5. The parameters of model identified in Chapter 5 provide the basic input data for the simulation. All the calculations are implemented in Cast3M software.

6.4.1 Phase transformation and transformation plasticity verification

The first case is devoted to calculate phase transformation and transformation plasticity in one element in 3-D (QUA8, 8 nodes). The algorithms of phase transformation and transformation plastic models are respectively given in section 6.2 .1 and section 6.2.3. It is supposed that the austenitic transformation is guided by linear law as we mentioned in section 6.2.1. We use the parameters of martensitic phase transformation model (Koistinen and Marburger): $M_s = 200$, $A_{c_1} = 700$, $A_{c_2} = 800$; and the parameters of transformation plasticity model (Leblond): $\sigma_\gamma^y =$ 217MPa, $\Delta\varepsilon_{\alpha-\gamma}^{th} = 9.58\mathrm{E}-03$. The thermal loading is linear heating and cooling with time, and the highest temperature reaches 1000°C (Figure 6.2).

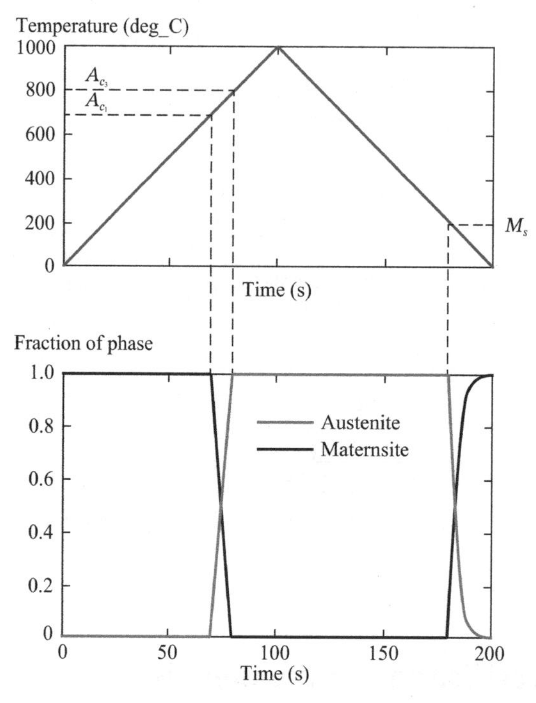

Figure 6.2 Fraction of phase evolution under temperature loading

The calculated phase fraction is given in Figure 6.2. It is in a good agreement with the experimental results in Chapter 5. The transformation plasticity is calculated in the situation of various applied stresses. Figure 6.3 gives the plastic strain induced by transformation. Total strains including transformation plasticity in experiment and simulation are shown in Figure 6.4. The numerical results fit the experimental data. The experimental strain increases slower than the simulated from 600°C to 800°C during heating because the precipitation hardening occurs in this temperture region and this factor is not taken into account. The martensite start temperature (M_s) in simulation is also different from that in experiment. The experimental study shows the martensite start temperature is affected by the highest temperature (see Section 5.7.1.). In our calculation, M_s is supposed a constant not a function of highest temperature for simplicity and also by lack of sufficient information.

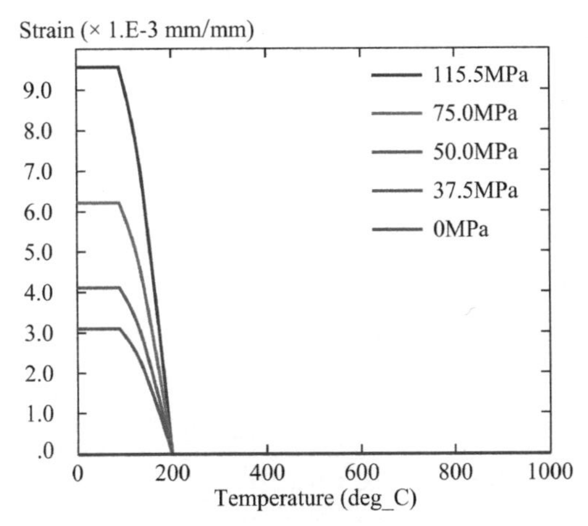

Figure 6.3 Transformation plasticity

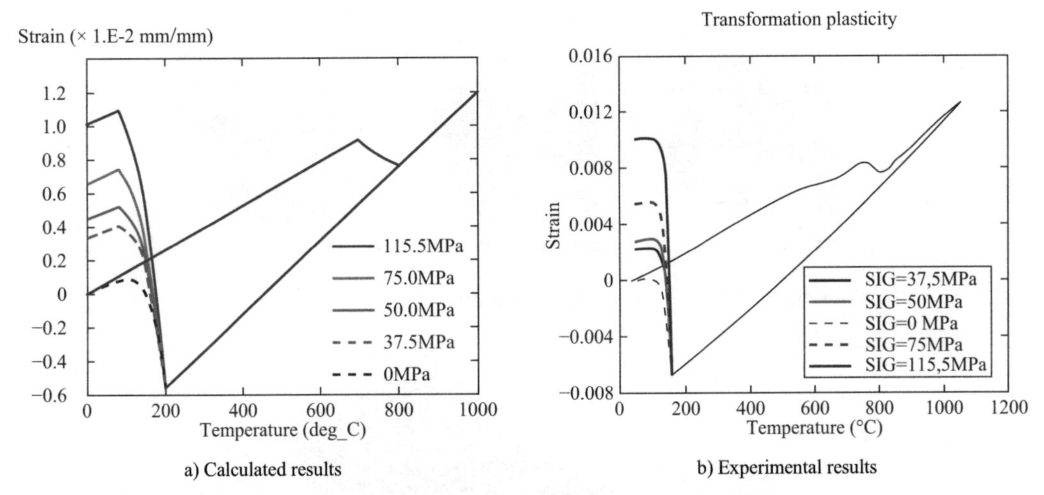

a) Calculated results

b) Experimental results

Figure 6.4 Total strain including transformation plasticity (calculated data vs. experimental data)

6.4.2 Numerical verification of flat notched bars

In this section, the study is devoted to strain, triaxial stress and damage of notched specimen. The first example is the comparison of calculated results of three cases with various notch radii (Case A R = 1.0mm, Case B R = 2.5mm, Case C R = 4.0mm), which are tensile tests at room temperature. And then these numerical results will compare to experimental data (F1A, F1B, F1C). Only one fourth of the flat specimen is taken into account because of the symmetry in 2-D calculation (Figure 6.5). The mesh consisted of 2100 QUA4-type elements. The boundary condition and geometry are shown in Figure 6.5. In simulation of triaxial stress state and damage, the method of plane stress, finite strain and displacement is preformed. In 3-D calculation, one eighth of whole specimen (flat part) is modelled with 600 CUB8-type (8-node volume) elements.

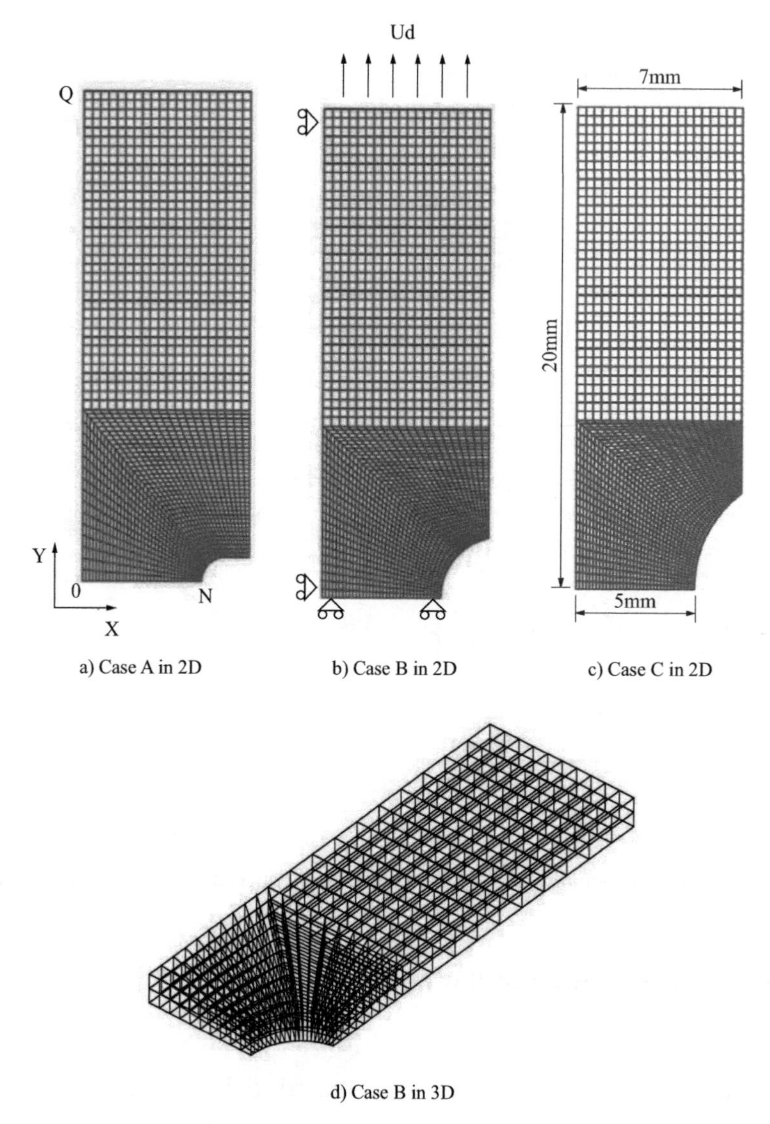

a) Case A in 2D b) Case B in 2D c) Case C in 2D

d) Case B in 3D

Figure 6.5 Meshes of notched specimens

Since there are experimental result of tensile test of notched specimens (CaseA) The triaxial stress state of three cases at minimum section of specimens (line ON) is displayed in Figure 6.6. It is shown numerical result of longitudinal strain (EPYY) is in a good agreement with experimental data in the figure: maximum strain is at edge of notch, and the specimen with large notch has flat strain distribution at minimum section. Figure 6.7 gives strain EPYY at longitudinal section (line OQ) with both experimental and simulated results.

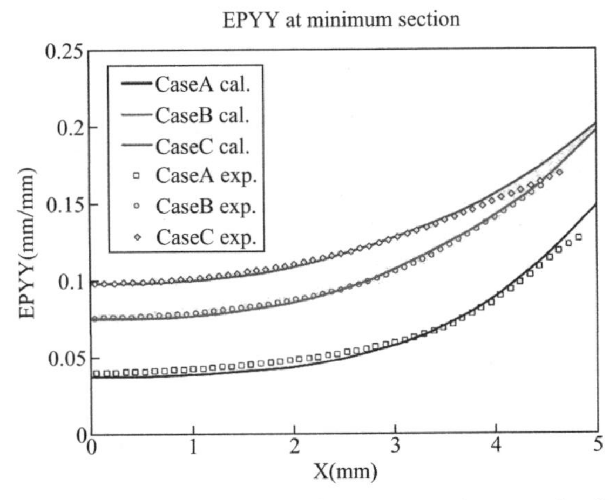

Figure 6.6 Strain (EPYY) at minimum section (ON) when displacement loading is equal to 1.0mm

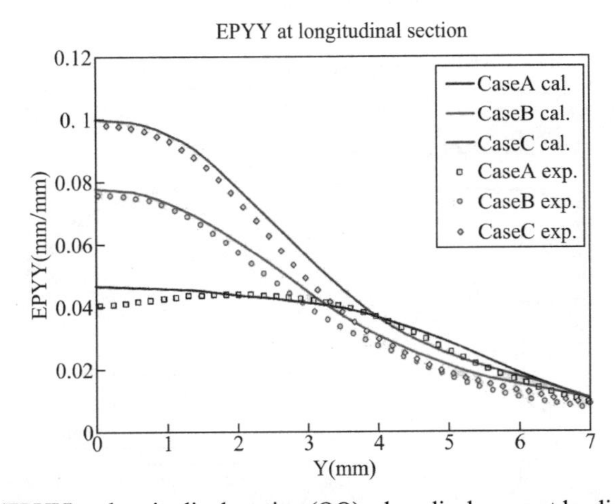

Figure 6.7 Strain (EPYY) at longitudinal section (OQ) when displacement loading is equal to 1.0mm

Stress triaxiality distribution across the minimum section specimen does not only depend on applied displacement but also on notch radius. Figure 6.8 shows that the notch has important an influence the distribution of triaxial stress: the maximum triaxial stress of Case B and Case C is at centre whereas it for Case A is near edge of notch. Stress triaxiality increases slightly with the increasing of loaded displacement in Figure 6.9.

In Figure 6.10, the evolution of the stress triaxiality of Case B with plastic strain is given for three locations: the centre core of the minimum section (O), the point at edge of notch (N) and middle point between both (L). This curves in Figure 6.10 show that stress triaxiality in three points remains constant at first because of elastic stage, and then increase with the evolution of plasticity in Case B and C, after then slightly drops in correspondence with the necking of specimen. However, the stress triaxiality at centre (O) is hardly influenced by plastic strain.

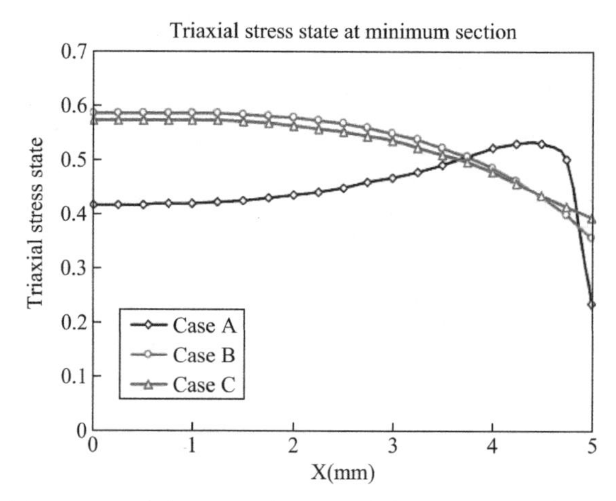

Figure 6.8 Triaxial stresses at minimum section with different notches (Case A, B, C) when loaded displacement is equal to 1.0mm

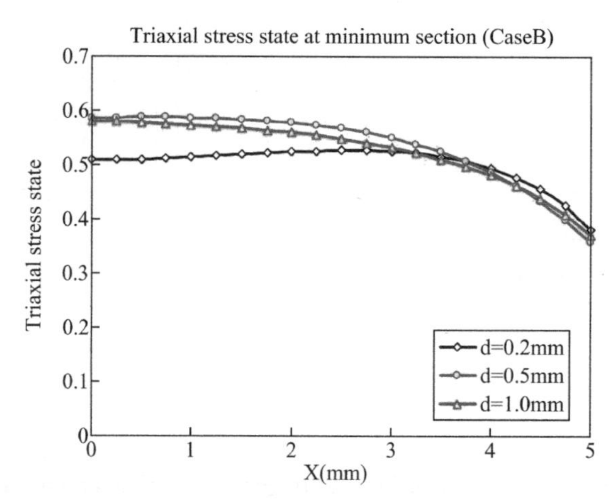

Figure 6.9 Triaxial stresses at minimum section of Case B with different applied displacement (d = 0.2, 0.5, 1.0mm) on specimen

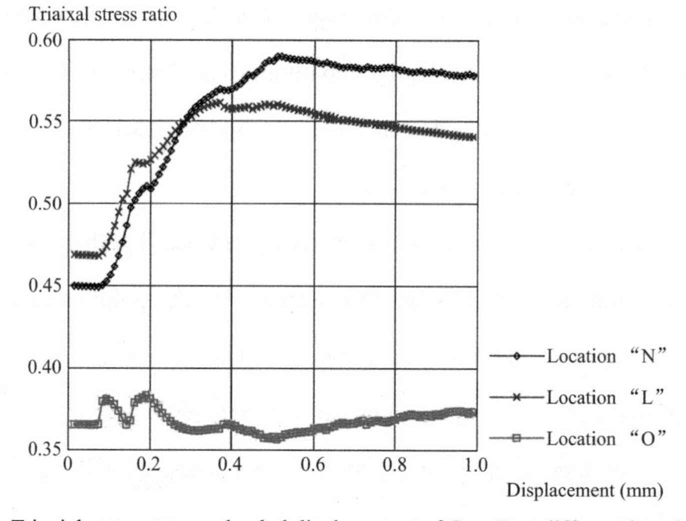

Figure 6.10 Triaxial stress state vs. loaded displacement of Case B at different locations (N, L, O)

The damage evolution across the minimum section has been analyzed in order to determine the first ductile failure location Damage simulation in notched geometry has been obtained analytically from the constitutive equations proposed in Chapter 4. The Equation 4-20 shows the damage evolution is a function of triaxial stress and plastic strain.

In Figure 6.11, damage contour map shows ductile failure evolution across the minimum section. The initial failure of Case A is near edge of notch whereas the maximum damage of Case C is at the centre of specimen. Compared with experimental data obtained in Chapter 5, the location of initial failure from simulated results is in a good agreement. The damage is mainly influenced by hydraulic stress. Here, the competition of stress triaxiality and plastic strain accumulation, which are controlled by the notch effect, determines failure initiation location (near the notch in Case A). Since the initial failure occurred, failure extends across the minimum section and finally entire failure happens. Consequently, the flat notched specimen can be considered an appropriate geometry for the study of damage evolution under multiaxial stress state.

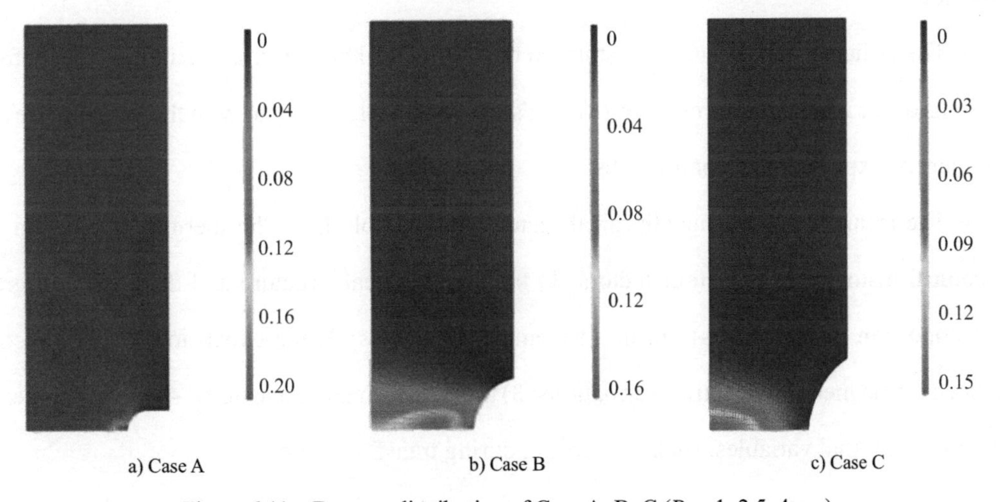

Figure 6.11 Damage distribution of Case A, B, C (R = 1, 2.5, 4mm)

6.5 Simulation of a disk heated by laser

6.5.1 Introduction

The introduction of welding join was introduced in Section 3.5.2 in detail. In order to simplify the problem and focus on the metallurgical and mechanical analysis, an example of disk heated by laser will be provided. Compared the welding joint (Figure 3.4), the case of disk is applied to simulate the HAZ and base metal, and the molten weld part is neglected (Figure 6.12). In the section, we will simulate the case of disk heated by laser for purpose of understanding and analyzing the damages and residual stresses produced during a welding operation.

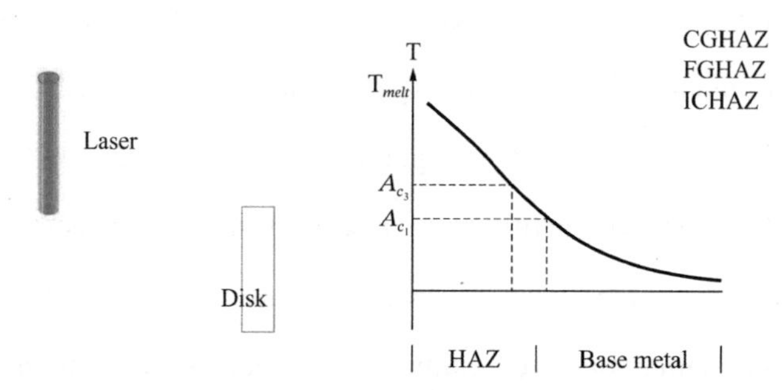

Figure 6.12 Schematic diagram of disk heated by laser and its HAZ[5]

In the case of simulation, the influences of thermal, metallurgical and mechanical aspects can be included as:

➢ The influence of the thermal history on the mechanical history results simultaneously from variations of the mechanical properties (Young′s modulus, yield strength) with the temperature and from thermal expansions or contractions.

➢ The influence of the metallurgical history (inseparable from the thermal history) on the mechanical history results in four factors: 1) the metallurgical structure and the austenitic grain size (proportion of the phases) on the mechanical properties; 2) the expansion and contraction produced by the metallurgical transformations; 3) the transformation plasticity; 4) the phenomenon of restoring internal variables, such as damage, during transformations.

The phenomenon of damage memory during martensitic transformation can be modelled through introduction of a memory coefficient η. The memory effect is null if $\eta = 0$ and full if $\eta = 1$. For this disk simulation, the memory coefficient can be considered to be null.

This simulation is performed in code Cast3M. The structure of simulation program is shown in Figure 6.13. Firstly the temperature filed is generated and then we have grain size and phase fraction fields. They are input files of main program, which consists of two

5 Coarse grained heat affected zone (CGHAZ): $T_{max} \gg A_{c_3}$. Fine g ained heat affected zone (FGHAZ): T_{max} is just above A_{c_3}. Inter critical heat affected zone (ICHAZ): $A_{c_1} < T_{max} < A_{c_3}$.

subroutines for calculation of damage and TRIP. Finally, the calculated results are transferred to post process.

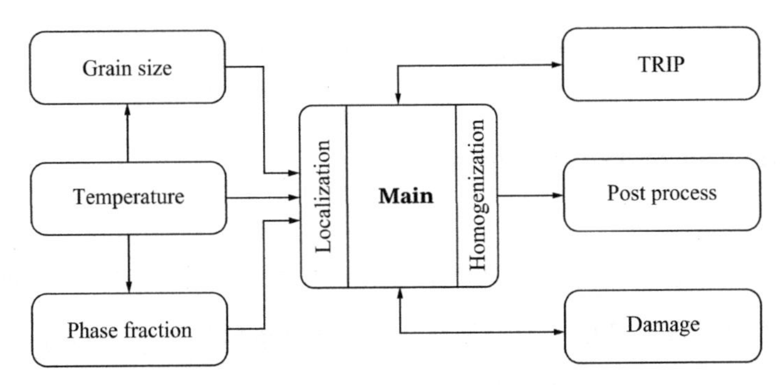

Figure 6.13 Structure of the program

6.5.2 Finite elememt simulation

A disk made of 15-5PH with 160mm in diameter and 5mm in thickness, was heated at the centre by a spot laser for 70 seconds, then cooled by natural convection for 730 seconds. The spot laser was modelled by a heat flux whose distribution on the upper side (Figure 6.14). The lower and lateral sides were subjected to free convection. The coefficient of exchange convection is: $h_c = 10\text{Wm}^{-2} \cdot {}^{\circ}\text{C}^{-1}$. The radiation coefficient is expressed by $h_r = \sigma\varepsilon$, where the emissivity: $\varepsilon = 0.7$ and the Helmotz constant: $\sigma = 5.67 \times 10^{-8}$. The ambient temperature is supposed to 20°C. The mesh consisted of 200 QUA4-type elements and the problem is considered axisymmetric (Figure 6.15).

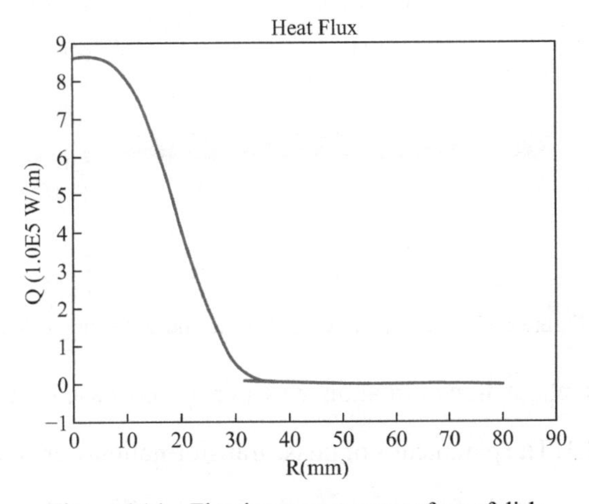

Figure 6.14 Flux input on upper surface of disk

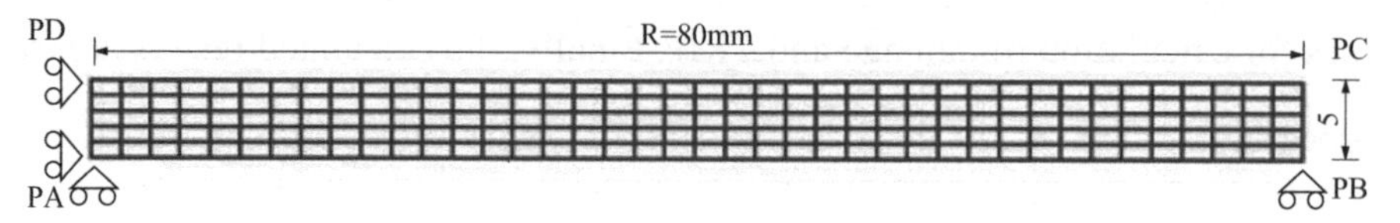

Figure 6.15 Mesh and dimensions of disk

6.5.3 Thermal and metallurgical results

The time for heating is 70 seconds and the cooling time is until to 800 seconds. Temperature evolution at location "PA", "PB" and "PD" is shown in Figure 6.16. The temperature filed at the moment of heating finish is observed in Figure 6.17. During the heating stage, the highest temperature reaches to 1050°C, and the temperature drops to 60°C at the end of cooling.

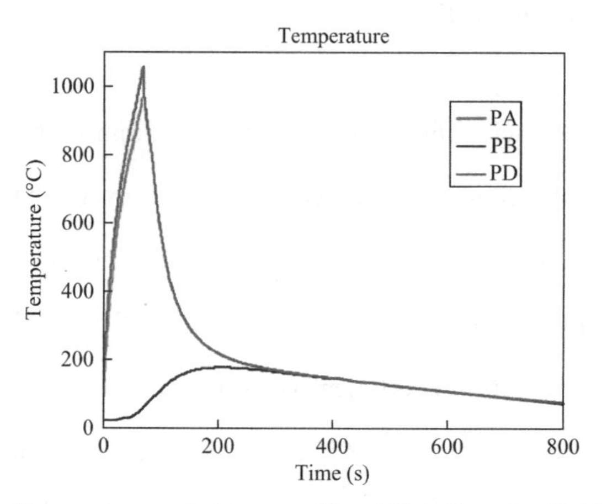

Figure 6.16 Temperature evolution on surface of disk (Location "PA" "PB" "PD")

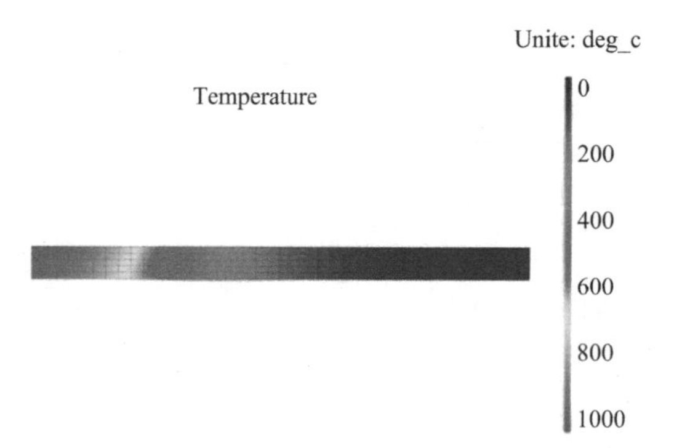

Figure 6.17 Temperature field at the end of heating (70 s)

The simulation of phase transformation was performed based on the calculation method presented in Section 6.2.1. The parameters of phase transformation were identified in Section 5.7.3. The phase proportion at the end of heating stage is plotted in Figure 6.18. It shows that the phase at the center (CGHAZ and FGHAZ) is austenite and it is mixture of martensite and austenite at the ICHAZ. The calculation of grain size was implemented according to the method presented in Section 6.2.2. Figure 6.19 shows the grain size distribution in disk at the end of cooling. One may observe a large gradient of grain size in the heated zone.

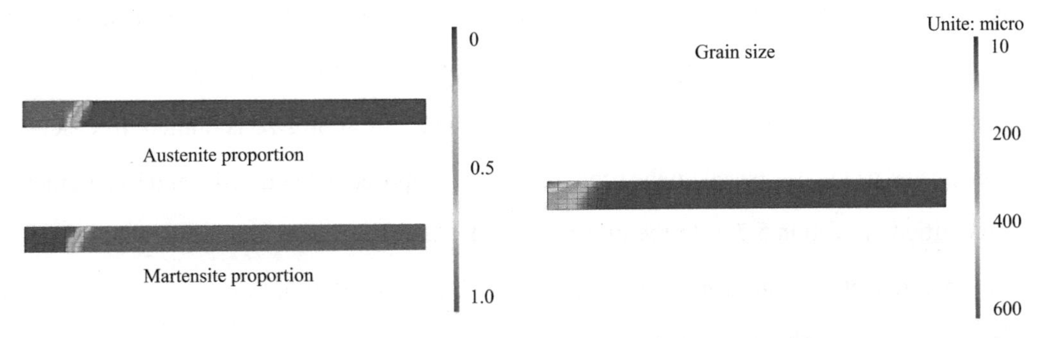

Figure 6.18 Phase proportion at the end of heating (70 s)

Figure 6.19 Grain size at the end of cooling (800 s)

6.5.4 Mechanical results

The mechanical simulation used the previously calculated results: temperature, phase proportion, and grain size. The mechanical calculation is not coupled with the temperature simulation, and the calculated temperature filed is an input date file. Phase fraction is determined by thermal loading not by mechanical loading. Transformation plasticity is coupled by mechanical condition. The time discretization can be different between the thermometallurgical and the mechanical calculations. The material properties are nonlinear with temperature, and experimental data at some specific temperatures are input in the program (see Chapter 5 for material data). Other values between the specific temperatures are interpolated linearly.

Various calculations were performed: the first one applied macro elastoplastic model with the mixing law of material properties (namely Calcul_1). The second calculation (Calcul_2) used the meso elastoplastic model in which each phase has its own material property, but here the transformation plasticity was not taken into account. Compared with Calcul_2, the Calcul_3 considered the transformation plasticity and applied two-scale model presented in Section 4.5. The last simulation (Calcul_4) used the two-scale damage model and introduced all factors: phase transformation, transformation plasticity, damage, grain size. The detail conditions of calculation are given in Table 6.1. All calculations except the first one took into account the affect of grain size.

The used mechanical properties of 15-5PH are given in the appendix B. The damage parameters are given in Section 5.7.4. The temperature filed, phase fraction and grain size were calculated in Section 6.5.3. As far as the affect of grain size is concerned, it was carried out through changing the yield stress using the Hall-Petch law (see Section 2.4.2). This effect is focused on the center of disk associated with highest temperature. The zone is not the region of high stress. Therefore, for the case of disk heated by laser, the influence of grain size is limited. In Calcul_4, the damage was calculated based on the proposed model proposed in Section 4.3 and its parameters were identified in Section 5.7.4. In the two-scale model, an important parameter should be noted: memory coefficient η. It indicates how internal variables in mother phase are transferred to the daughter phase when phase transformation occurs. In our calculation, it is supposed to be zero because it is not easy to choose a specific value.

Comparison of calculational condition of four simulations Table 6.1

No.	Hardening	Phase change	Grain size	TRIP	Damage
Calcul_1	Kinematic	NO	NO	NO	NO
Calcul_2	Kinematic	YES	YES	NO	NO
Calcul_3	Kinematic + isotropic	YES	YES	YES	NO
Calcul_4	Kinematic + isotropic	YES	YES	YES	YES

Figure 6.20 shows typical deformation (magnification 10X) at the end of heating and at the end of cooling: the highest displacement is at the centre of disk when it is at the end of heating, whereas the highest deformation is at the zone (ICHAZ[6], see Section 3.5.2) where partial transformation occurs at the end of cooling. The ICHAZ is also a region where stresses concentrate.

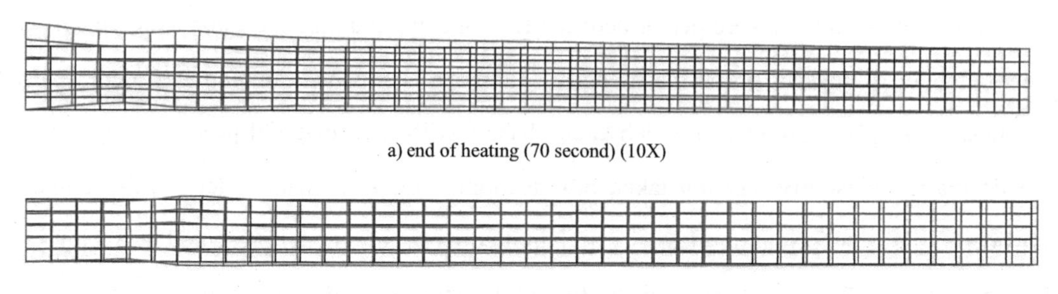

a) end of heating (70 second) (10X)

b) end of cooling (800 second) (10X)

Figure 6.20 Deformed shape of disk (Calcul_4)

6 Inter critical heat affected zone (ICHAZ): $A_{c_1} < T_{max} < A_{c_3}$.

Figure 6.21 shows typical plastic strain in radial direction (EPRR) and the plastic strain in circumferential direction (EPTT) in macro scale. The disk is stretched in both radial and circumferential direction at the centre of disk, and is compressed near ICHAZ. The strains and stresses are concentrated in the HAZ.

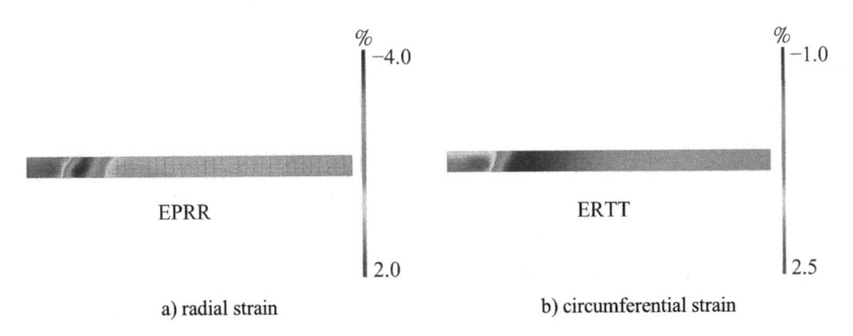

a) radial strain b) circumferential strain

Figure 6.21 Macro plastic strain distribution of disk at the end of cooling (Calcul_4)

From Figure 6.22 to Figure 6.25, typical stresses on the upper surface of disk are respectively calculated in four simulation (from Calcul_1 to Calcul_4). Compared to the other three calculations, the Calcul_1 with macro elastoplastic model has smooth stress distribution on the surface and higher maximum stress (Figure 6.22). The meso model with two phases not only changes the stress distribution but also lower maximum stress in radial and circumferential direction. When the transformation plasticity is taken into account (Calcul_3), the maximum stresses are decreased to some extent. The peak stresses distribute similarly on upper surface of disk in the last three calculations: the positive peak stress is at the ICHAZ, whereas the negative lies in FGHAZ near ICHAZ side.

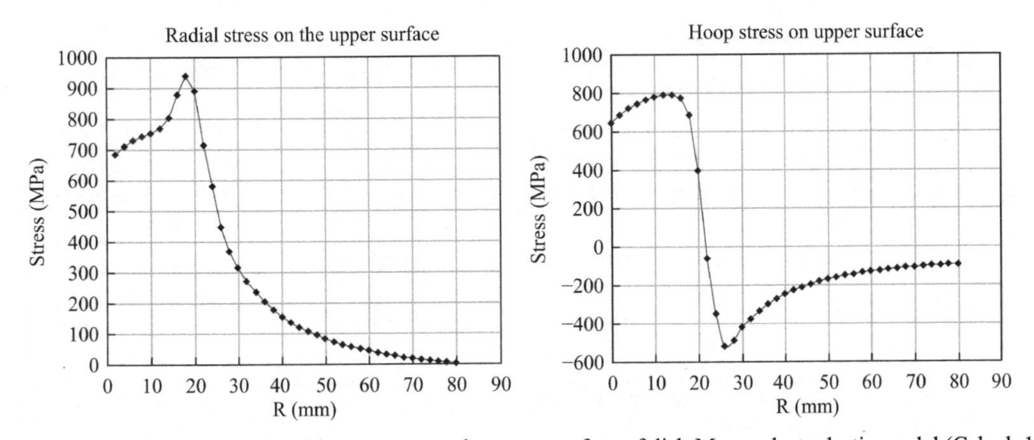

Figure 6.22 Residual radial and hoop stress on the upper surface of disk-Macro elastoplastic model (Calcul_1)

The calculation result shows that damage has influence on residual stresses. Comparing

Calcul_3 and Calcul_4, the damage reduces the maximum radial stress of about 10 percent, and 8 percent for circumferential stress.

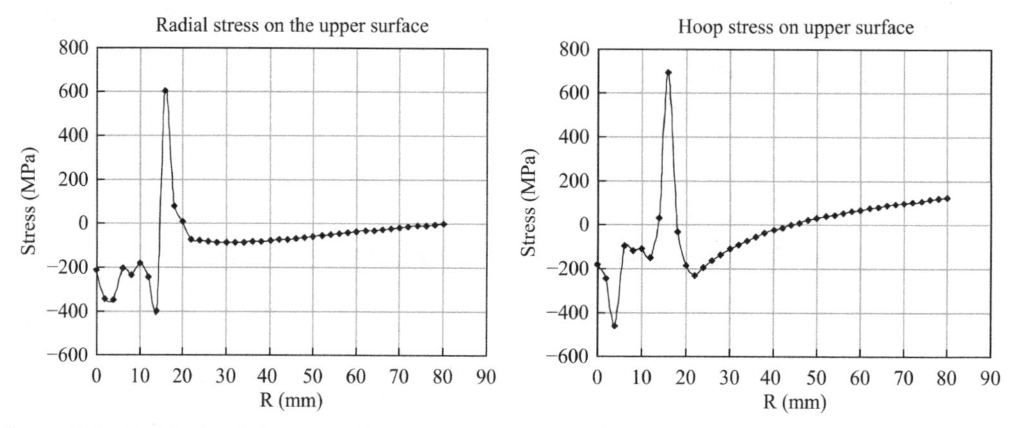

Figure 6.23 Residual radial stress and hoop stress on upper surface of disk-Meso elastoplastic model without TRIP (Calcul_2)

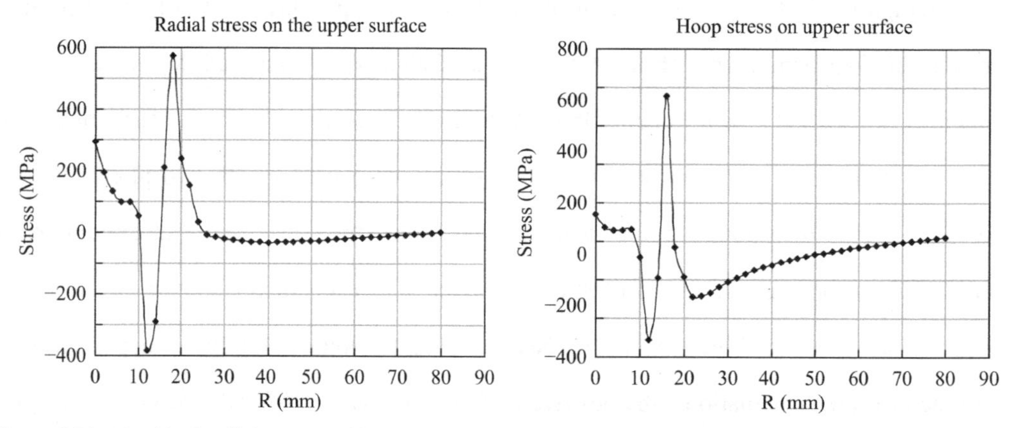

Figure 6.24 Residual radial stress and hoop stress on upper surface of disk-Two-scale elastoplastic model with TRIP (Calcul_3)

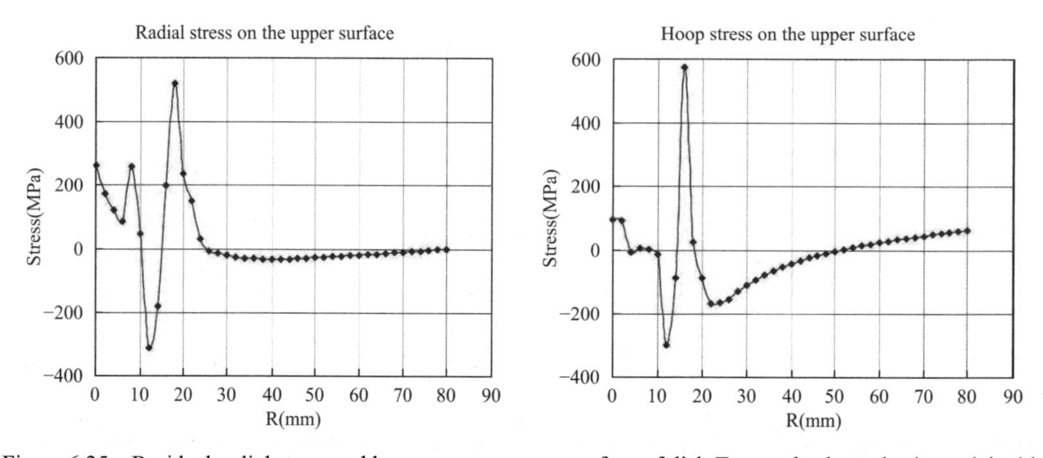

Figure 6.25 Residual radial stress and hoop stress on upper surface of disk-Two-scale elastoplastic model with TRIP and damage (Calcul_4)

Table 6.1 shows the comparison of minimum and maximum stress among four calculations. It indicated that the TRIP and damage decrease the peak stresses in radial and circumferential direction. The complete simulation (Calcul_4) shows the residual stress of disk heat by laser in radial direction reaches up 519 MPa and the peak stress in circumferential direction is up to 574MPa.

The damage was plotted at three representative moments: at 50 seconds when austenitic transformation does not happen (Figure 6.26), at 100 seconds when austenite occurs at the centre of disk (Figure 6.27), and at 800 seconds at which the martensitic transformation is completed (Figure 6.28). The damage has happened before austenite occurred because of heating induced expansion at centre and restriction of surrounding cold metal. Even though the value of damage is very small, it indicated damage occurs firstly at centre. However, the damage was not observed at the centre in Figure 6.27 because the central metal was changed into austenite and damage did not inherit from martensite due to null memory coefficient η. At that moment, damage appeared on the boundary between martensite and austenite. When martensitic phase transformation was completed with the decreasing of temperature, damage occurred and accumulated in the whole heat affected zone.

Comparison of stress on the upper surface of disk (MPa) Table 6.2

No.	Model	σ_r^{max}	σ_r^{min}	σ_θ^{max}	σ_θ^{min}
Calcul_1	macro	907	0	791	0
Calcul_2	meso elastoplastic	601	−396	692	−461
Calcul_3	2scale + TRIP	570	−384	610	−336
Calcul_4	2scale + TRIP + damage	519	−312	574	−301

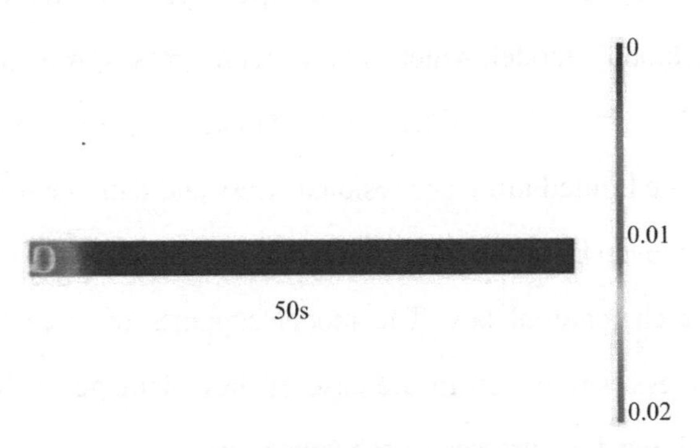

Figure 6.26 Damage distribution of disk at 50 second

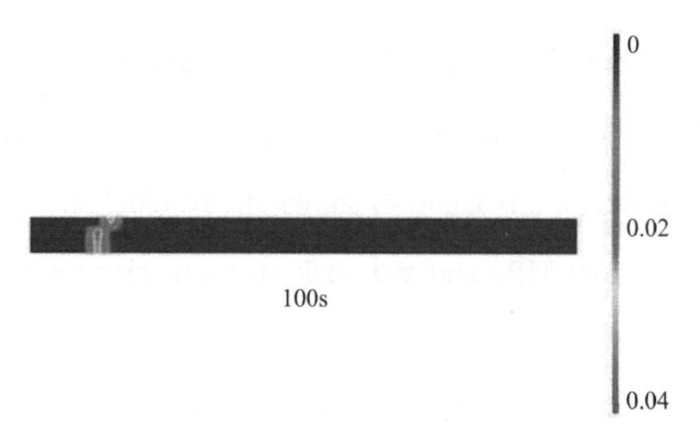

Figure 6.27 Damage distribution of disk at 100 second

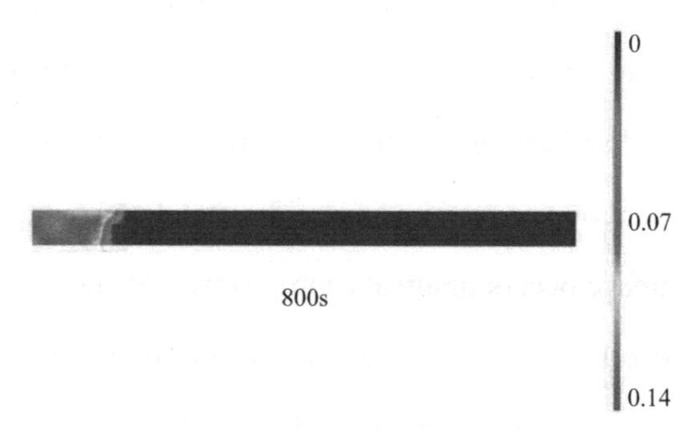

Figure 6.28 Damage distribution of disk at 800 second (end of cooling)

6.6 Conclusion

Numerical calculation of the model presented in Chapter 5 was implemented in software Cast3M in the chapter. Numerical verification of flat notched bars showed a good agreement with the experimental results in term of strain and damage on the surface. The example of disk heated by laser was used to simulate the HAZ of welding.

In metallurgic aspect, the Koistinen and Marburger's empirical law which does couple mechanical behaviour has a good agreement with the experimental results. Transformation plasticity behaviour guided by Leblond's model, which couples with stresses, was quantified and introduced in the calculation. As plasticity, TRIP influenced the damage to some extent and affected residual stresses. Grain size has the limited influence residual stress and damage in the case of heated disk.

As far as the mechanical calculation is concerned, the proposed two-scale model brings the freedom of choosing each material law. The model supports to trace history of each phase's behaviours (damage, stress, strain …). In the case of disk, damage in the martensite decreased about 10 percent of peak residual stresses on the upper surface.

Chapter 7 Conclusions and Perspectives

The simulation of thermomechanical problems with phase transformation is a formidable problem which many research teams dedicated to. When the thermomechanical problems couple damage, the problems become more complicated. The difficulty lies not only in the coupling between phase transformation (almost accompanying transformation plasticity) and mechanics but also in that damage develops in the multiphase materials.

Our contribution was devoted to three particular points. The first objective was to develop proper constitutive equations that can correctly predict damage and residual stresses induced by severe thermal or mechanical loading. The constitutive equations included phase transformation, transformation plasticity and damage models. The existed phase transformation (Koistinen and Marburger'law) and transformation plasticity models (Leblond's model) which we chosen were satisfied in the simulation of material 15-5PH. A damage equation was proposed based on the Lemaitre's ductile damage model for the purpose of extending application and better fitting our studied material. The constitutive equations used a two-scale structure which provides a freedom to choose material law for each phase and where all the phases are represented with their own behaviour.

The second aim was to provide an experimental database including thermal, metallurgic and mechanical properties of the stainless steel 15-5PH. The identification of phase transformation model, transformation plasticity model and damage model, besides the characteristics of material, were performed in the laboratory LaMCoS of INSA in France. The particular contribution was to study damage's mechanism, evolution and characterization. Tensile tests of round specimens were used to identify the parameters of damage model as well as mechanical behaviours at various temperatures. Tests of flat notched specimen were designed to provide the verification of damage model and strain localization due to the fact that various notch radii induce different triaxial stress ratio profiles in the mid section. The tests of flat notched specimen applied three dimensional image correlation technologies to trace the strain localization and evolution on surface of specimen. In our experimental study, our tests about phase transformation and damage were focused on the proportional loading in one dimension and did not take non proportional loading into account. Many test results showed variable and rational, even though the damage parameters at high temperature are less accuracy because the measurement of Young's module was difficult at over 800°C. The experimental database of 15-5PH provided the necessary parameters for the numerical calculation of residual stresses including phase transformation, transformation plasticity and damage. In addition, the microscopic experiment (XRD, TEM and SEM) were performed to provide microstructure characterization of 15-5PH and discover the damage mechanism.

The third objective was to implement numerical calculation in the code Cast3M. On the one hand, numerical verification of the flat notched bars was performed to compare with experimental results. The results showed the numerical result has a good agreement with the experimental data. On the other hand, we used the two-scale model including phase transformation, transformation plasticity and damage to simulate the level of residual stresses of the disk made of metal 15-5PH heated by laser. The internal variables, such as strain, stress, damage, were successfully traced in the simulation of two-scale model. The simulation results showed the transformation plasticity changes the level of residual stresses and is not negligible; damage decreases about 8 percent peak value of residual stresses on upper surface of disk. Even though the influence of grain size is obvious in austenitic steel such as 316L, it is small in our case because the coarsening of austenite is concentrated at the centre of disk where the stress is not high. Furthermore, the relationship between austenite grain size and the formation of martensite is not clear.

The thermomechanical problems coupling with damage in the phase-change metal are really difficult. Our experimental study leaves some problems to be further studied in the future. As far as the damage model and simulation are concerned, there are also still some problems which are open.

In experimental aspect:

➢ The microscopic test of specimen during phase transformation is too difficult to realize because the time of phase transformation is extremely short.

➢ The microscopic observation at high temperature is also a problem because of the oxidation on the surface of specimen.

➢ It is verified that the martensite start temperature (M_s) is influenced by maximum temperature reached. The accurate relationship between both needs more experiments to explore it.

➢ When we use two cameras to trace the evolution of traction of notched specimen, the image correlation provides three dimensional displacement of surface but lacks of strain in thickness direction because the monitoring on only one side surface can not provides the thickness data of specimen. Seeing that the three dimensional strain is important to study strain localization and damage, a possible solution is to use four cameras (setting two on each side of specimen) to trace both sides and then generate strains in three direction through image correlation.

In modeling aspect:

➢ Is there an interaction between damage in austenite and damage in martensite? If there is, how do they affect with each other?

➢ Do grain size and transformation plasticity influence damage or vice versa? Does the memory effect exist during martensite and austenite transfer to each other (memory coefficient $\eta \neq 0$)?

➢ It was known the precipitation is an important phenomenon in stainless steel 15-5PH, it seems more reasonable to take the behaviour into our modelling.

➢ In the simulation of residual stress, another important factor is hydrogen diffusion which may be important for damage and crack.

In numerical aspect:

➢ The model will be applied to calculate complicated structure or weld joint.

Bibliography

[1] 7th International Conference on Trends in Welding Research Conference, Georgia, USA, May 16-20, 2005.

[2] 8th International Seminar on the Numerical Analysis of Weldability, Graz-Seggau, Institute of Materials, Austria, 2006.

[3] 11th Modeling of Casting, Welding and Advanced Solidification Processes, Opio, France, 2006.

[4] Abrassart F., Influence des transformations martensitiques sur les propriétés mécaniques des alliages du système Fe-Ni-Cr-C, Thèse de Doctorat ès-Sciences Physiques, Université de Nancy I, 1972.

[5] Acharya A., Bassani J.L., Lattice incompatibility and a gradient theory of crystal plasticity, Journal of the Mechanics and Physics of Solids, 48:1565-1595, 2000.

[6] Alberry P.J., Jones W.K.C., Marchwood Engineering Lab., Internal Report No. R/M/R282, 1979.

[7] Aliaga C., Massoni E., and Treuil J.L., 3d numerical simulation of thermo-elastovisco-plastic behaviour using stabilized mixed f.e. formulation: application to 3d heat treatment, In Ⅳ World Congress on Computational Mechanics, Buenos Aires, Argentina, June-July 1998.

[8] Aliaga C., Massoni E., Louin J.-C., and Denis S., 3D finite element simulation of residual stresses and distortions of cooling work pieces. In: Third International Conference on Quenching and Control of Distortion. Prague, Czech Republic, 1999.

[9] Andersson B.A.B., Thermal stresses in a submerged-arc welded joint considering phase transformations, ASME J. Engrg. Mater. Technol. 100: 356-362, 1978

[10] Avrami M., Kinetics of phase change i: general theory, Journal of Chemical Physics, 7: 103-112, 1939.

[11] Avrami M., Kinetics of phase change ii: transformation-time relations for random distribution of nuclei. Journal of Chemical Physics, 8: 212-224, 1940.

[12] Badeshia H.K.D.H., Bainite of steels, The Inst. of Meter, Ed., 1992.

[13] Bammann D.J., Mosher D., Hughes D.A., Moody N.R., and Dawson, P.R., Using spatial gradients to model localization phenomena, Sandia National Laboratories Report, SAND99-8588, Albuquerque, CA, 1999.

[14] Barnes A.M., Solidification cracking susceptibility of modified 9Cr1Mo submerged arc weld metals: the influence of Mn and Nb, Proceedings of the 5th International Charles Parsons Turbine Conference, 407430. 2000.

[15] Berveiller M., and Fischer F.D. (Eds.), CISM Courses and lectures 368 Mechanics of Solids with Phase Changes. Springer, New York, 1997.

[16] Betten J., Damage tensors in continuum mechanics, J. Mec. Theor. Appl., 2: 13-32, 1983.

[17] Bigoni D., Piccolroaz A., Yield criteria for quasibrittle and frictional materials, Int. J. Solids Struct, 41: 2855-2878, 2003.

[18] Bonora N., A nonlinear CDM model for ductile failure, Engineering Fracture Mechanics, 58: 11-28, 1997.

[19] Bonora N., On the effect of triaxial state of stress on ductility using nonlinear CDM Model, International Journal of Fracture, 88: 359-371, 1998.

[20] Bonora N., Ruggiero A., Milella P.P., Fracture energy effect on spall signal, In: Proceedings of the APS conference on shock compression of condensed matter, Portland, USA, 2003.

[21] Briottet L., Martinez M., and Burlet H., Introduction dans castem 2000 d'un module thermométallurgique, d'un module de diffusion et des couplages associés, Technical Report DEM n 98/09, CEA/LSMM, 1998.

[22] Bruhns O.T., Schiesse P., A continuum model of elastic-plastic materials with anisotropic damageby oriented microvoids, Eur. J. Mech., A/Solids, 15: 367-396, 1996.

[23] Brunet M., Morestin F., Experimental and analytical necking studies of anisotropic sheet metals, Journal of Materials Processing Technology, 112: 214-216, 2001.

[24] Brunet M., Morestin F., and Walter H., Failure prediction in anisotropic sheet-metals under forming operations using damage theory, Int. J. Form. Processes 5: 225-235, 2002.

[25] Brunet M., Morestin F., and Walter H., Failure analysis of anisotropic sheet-metals using a non-local plastic damage model, Journal of Materials Processing Technology, 170(1-2): 457-470, 2005.

[26] Brunig M., An anisotropic ductile damage model based on irreversible thermodynamics, International Journal of Plasticity, 19: 1679-1713, 2003.

[27] Cast3M is a computer code, developed by French CEA, for the analysis of structures by the finit element method (FEM) and the Computational Fluids Dynamics.

[28] Cavallo N. Contribution à la validation expérimentale de modèles décrivant la ZAT lors d'une opération de soudage, Thèse de Doctorat, INSA de Lyon, 1998.

[29] Chaboche J.L., Description thermodynamique et phénoménologique de la viscoplasticité cyclique avec endommagement, Thèse de Doctorat Es-Science, Paris, Ⅵ, 1978.

[30] Chaboche J.L., Anisotropic creep damage in the framework of the continuum damage mechanics, Nuclear Engineering and Design, 79:309-319, 1984.

[31] Chandrakanth S., Pandey P.C., A new ductile damage evolution model, International Journal of Fracture, 60: R73-R76, 1993.

[32] Cherkaoui M., Berveiller M., and Sabar H., Micromechanical modeling of martensitic transformation induced plasticity (TRIP) in austenitic single crystals, Int. J. Plasticity 14: 597-626, 1998.

[33] Cherkaoui M., Berveiller M., and Lemoine X., Coupling between plasticity and martensitic phase transformation: overall behaviour of polycrystalline TRIP steels, Int. J. Plasticity 16: 1215-1241, 2000.

[34] Cherkaoui M., Berveiller M., Sabar, H., Micromechanical modeling of martensitic transformation induced plasticity (TRIP) in austénitic single crystals, International Journal of plasticity, 14(7): 597-626, 1998.

[35] Cherouat A., Saanouni K., Hammi Y., Improvement of forging process of a 3D complex part with respect to damage occurrence, Journal of Materials Processing Technology, 142: 307-317, 2003.

[36] Chinese Mechanical Engineering Society, Welding manual: Material welding, China Machine Press, 2004.

[37] Choi S., Shah S.P., Measurement of deformations on concrete subjected to compression using image correlation, Exp. Mech., 37: 307-313, 1997.

[38] Chow C.L., Wang J., An anisotropic theory of continuum damage mechanics for ductile fracture, Eng. Frac. Mech., 27: 547-558, 1987.

[39] Chu C.C., Needleman A., Void nucleation effects in biaxially stretched sheets, J. Eng. Mater. Technol, 102: 249-256, 1980.

[40] Clerc P., Mesure de champs de déplacements et de déformations par stéréovision et corrélation d'images numériques, Thèse de Doctorat, INSA de Lyon, 2002.

[41] Clerc P., Morestin., Brune M., Measurement of deformation fields using numerical images correlation, proceedings of the AMPT'99 and IMCI6, 3:1889-1897, 1999.

[42] Colette G., Sur le comportement thermomécanique des matériaux et des structures métalliques, Thèse de Docteur-Ingénieur, Institut National Polytechnique de Lorraine, Nancy, 1980.

[43] Colonna F., Chenot J.L., Wendenbaum J., Denis S., and Gautier E., On thermoelastic-viscoplastic analysis of cooling processes including phases changes, Jal. Mater. Proc. Tech., Vol 34: 525-532, 1992.

[44] Constant, A., Henry, G., and Charbonnier, J.C., Principes de bases des traitements thermiques, thermomécaniques et thermochimiques des aciers, PYC éd. 1992.

[45] Coret M., Etude experimentale et simulation de la plasticite de transformation et du comportement multiphase de l'acier de cuve 16MND5 sous chargement multiaxial anisotherme, Ph.D Thesis, LMTCachan, Paris, France, 2001.

[46] Coret M. and Combescure A., A Mesomodel for the Numerical Simulation of the Multiphasic Behavior of Materials under Anisothermal Loading, International Journal of Mechanical Sciences, 44: 1947-1963, 2002.

[47] Coret M., Calloch S. and Combescure A., Experimental study of the phase transformation plasticity of 16MND5 low carbon steel under multiaxial loading, Int. J. Plasticity, 18: 1707-1727, 2002.

[48] Coret M., Calloch S. and Combescure A., Experimental study of the phase transformation plasticity of 16MND5 low carbon steel induced by proportional and nonproportional biaxial loading paths, European Journal of Mechanics - A/Solids, 23: 823-842, 2004.

[49] Denis S., Gautier E., Beck G., and Simon A., Stress-phase-transformation interaction, basics principle, modelling, and calculation of internal stresses. Materials Science and Technology, 1: 805-14, 1985.

[50] Denis S., Gautier E., and Sjostrom S., Influence of stresses on the kinetics of pearlitic

transformation during continuous cooling, Acta. Metall., 35: 1621-1632, 1987.

[51] Denis S., Sjostrom S., and Simon A., Coupled temperature, stress, phase transformation calculation model, numerical illustration of the internal stresses evolution during cooling of a eutectoide carbon steel cylinder, Metall., 18A: 1203-1212, 1987.

[52] Denis S., Considering stress-phase transformation interactions in the calculation of heat treatment residual stresses, Jal. Ph., 6: 159-174, 1996.

[53] Depradeux L., Simulation numerique du doudage: acier 316L, Thèse de Doctorat, INSA de Lyon, 2004.

[54] Desalos, Y., Comportement dilatométrique et mécanique de l'austénite métastable d'un acier A533, IRSID, rapport n° 95349401 MET44, 1981.

[55] Devaux, J., Comportement plastique des aciers en cours de transformations de phases-Etude numérique des lois de mélange et de la plasticité de transformation. Rapport Systus International LDEW98/235, 1998.

[56] Diani J.M., Contribution à l'étude du comportement d'un acier présentant de la plasticité induite par transformation sous sollicitations rapides, Thèse, ENSMP, 1992.

[57] Diani J.M., Sabar H., and Berveller M., Micromechanical modelling of the transformation induced plasticity (TRIP) phenomenon in steels, Int. J. Engrg. Sci. 33: 1921-1934, 1995.

[58] Diani J.M., Soler M., Berveller M., and Abar H., Elasto-plastic micro-macro modelling of solid-solid phase transformation: application to transformation induced plasticity, J. Phys. 5, C2-507-C2-512, 1995.

[59] Diani J.M., Berveller M., and Sabar H., Incremental micromechanical modelling of the transformation induced plasticity, J. Phys. 6, C1-419-C1-427, 1996.

[60] Drabek T., Bohm H.J., Damage models for studying ductile matrix failure in composites, Computational Materials Science, 32 : 329-336, 2005.

[61] Drabek T., Bohm H.J., Micromechanical finite element analysis of metal matrix composites using nonlocal ductile failure models, Computational Materials Science, 37: 29-36, 2006.

[62] El-Ahmar W., Robustesse de la simulation numérique du soudage: acier 316L, Thèse de Doctorat, INSA de Lyon, 2007.

[63] El-Ahmar W., Jullien J.F., Gilles P., Reliability of hardening model to predict the welding residual stresses, 3rd International Conference on Thermal Process Modelling and Simulation,

Budapest, Hungry, 2006.

[64] Enakoutsa K., Leblond J.B., Perrin G., Numerical implementation and assessment of a phenomenological nonlocal model of ductile rupture, Comput. Methods Appl. Mech. Engrg, 196: 1946-1957, 2007.

[65] Fernandes F.B.M., Denis S., and Simon A., Mathematical model coupling phase transformation and temperature evolution during quenching of steels, Material science and technology, 1(10): 838-844, 1985.

[66] Fischer F.D., A micromechanical model for transformation plasticity in steels, Acta Metall. Mater. 38: 1535-1546, 1990.

[67] Fischer F.D., Transformation induced plasticity in triaxially loaded steel specimens subjected to a martensitic transformation, Eur. J. Mech. A/Solids 11: 233-244, 1992.

[68] Fischer F.D., Sun Q.-P., and Tanaka K., Transformation-induced plasticity (TRIP), Appl. Mech. Rev. 49: 317-364, 1996.

[69] Fischer F.D., Modelling and simulation of transformation induced plasticty in elasto-plastic materials. Mechanics of solids with phase changes, Berveiller M., Fischer F.D. Eds Wien New York Springer, 368: 189-237, 1997.

[70] Fischer F.D., Oberaigner E.R., Tanaka K., and Nishimura F., Transformation induced plasticity revised an updated formulation, Int. J. Solids Struct. 35: 2209-2227, 1998.

[71] Fischer F.D., Reisner G., Werner E., Tanaka K., Cailletaud G., and Antretter T., A new view on transformation induced plasticity (TrIP), Int. Jal. Plas., 16: 723-748, 2000.

[72] Fleck N.A., Muller G.M., Ashby M.F., and Hutchinson J.W., Strain gradient plasticity: theory and experiment, Acta Metallurgica et Materialia, 42: 475-487, 1994.

[73] Fleck N.A., Hutchinson J.W., Strain gradient plasticity, Advance in Applied Mechanics, 33: 295-361, 1997.

[74] Folkhard E., Welding Metallurgy of Stainless Steels. Springer-Verlag, Wien, New York, 1988.

[75] Friedman E., Thermomechanical analysis of the welding process using the finite element method, ASME J. Press. Vess. Technol. 97 3: 206-213, 1975.

[76] Gao X., Kim J., Modeling of ductile fracture: significance of void coalescence, Int. J. Solids Struct, 43: 6277-6293, 2006.

[77] Gautier E., Transformations perlitiques et martensitique sous contrainte de traction dans les

aciers, Thèse de Doctorat, Sciences Physiques, Institut National Polytechnique de Lorraine, Nancy, 1985.

[78] Gautier E., Denis S., Liebaut C., and S. Sjöström, Mechanical behaviour of fe-c alloy during phase transformations, Jal. Ph. Ⅳ, C3(4): 279-284, 1994.

[79] Gautier E., Interactions between stresses and diffusive phase transformation with plasticity, Mechanics of solids with phase changes, M. Berveiller and F.D. Fischer, Springerwiennewyork edition, 1997.

[80] Germain P., and Muller P., Introduction à la mécanique des milieux continus, Masson, 1980.

[81] Gologanu M., Leblond J.B., Devaux J., Approximate models for ductile metals containing nonspherical voids - case of axisymmetric prolate ellipsoidal cavities, J. Mech. Phys. Solids, 41: 1723-1754, 1993.

[82] Gologanu M., Leblond J.B., Devaux J., Approximate models for ductile metals containing nonspherical voids - Case of axisymmetric oblate ellipsoidal cavities, J. Eng. Mater. Tech., 116: 290-297, 1994.

[83] Gologanu M., Leblond J.B., Perrin G., Devaux J., Recent extensions of Gurson's model for porous ductile metals, In: Suquet, P. (Ed.), Continuum Micromechanics. Springer-Verlag, 61-130, 1995.

[84] Gooch T.G., Welding Martensitic Stainless Steels. Welding Institute Research Bulletin, 18: 343-349, 1977.

[85] Greenwood, G.W., and Johnson, R.H. The deformation of metals under small stresses during phase transformation, Proc Roy Soc, 283: 403-422, 1965.

[86] Grunwald, A fatigue model for shape optimization based on continuum madage mechanics, Thesis, University Karlsruhe, Germany, 1996.

[87] Gurson A.L., Continuum theory of ductile rupture by void nucleation and growth: Part I - Yiel criteria and flow rules for porous ductile media, J. Eng. Mat. Tech, 99: 2-15, 1977.

[88] Habraken A.M., and Bourdouxhe M., Coupled thermo-mechanical-metallurgical analysis during the cooling process of steel pieces, E. J. Mech. A./Solids, 11(3): 341-402, 1992.

[89] Hamata N., Billardon R., et al., A model for nodular graphite cast iron coupling an isothermal elastoviscoplasticity and phase transformation. In: Desai CS et al., editors. Constitutive laws for engineering materials. New York: ASME Press, 593-6. 1991.

[90] Hamata N., Modelisation du Couplage Entre L'elasto-viscoplasticite Ansotherme et la Transformation de Phase D'un Fontr G.S. Ferritique, These de Doctorat de L'universite Paris 6, 1992.

[91] Hambli R., Comparison between Lemaitre and Gurson damage models in crack growth simulation during blanking process, International Journal of Mechanical Sciences 43: 2769-2790, 2001.

[92] Hashin Z. ans Shtrikman S., A variational approach to the theory of the effective mangnetic permeability of multiphase materials, Jal. Applied Ohys., 33: 3125-3131, 1962.

[93] Helm J.D., McNeill S.R., and Sutton, M.A., Improved 3-D image correlation for surface displacement measurement, Opt. Eng., 35(7): 1911-1920, 1996

[94] Hibbitt H., and Marcal P., A numerical thermo-mechanical model for the welding and subsequent loading of a fabricated structure, Comput. Struct. 3: 1145-1174, 1973.

[95] Horikawa, H., Ichinose, S., Morii, K., Myazaki, S., Otsuka, K., orientation dependance of $\beta_1 \rightarrow \beta_1'$ stress induced martensitic transformation in a Cu-Al-Ni alloy. Metall. Trans., A19: 915-923, 1988.

[96] HTTP://www.hightempmetals.com

[97] Hutchings M.T., Krawitz A.D. (Eds.), Measurement of Residual and Applied Stress Using Neutron Diffraction, NATO ASI Series E No.216, Kluwer, Netherlands, 1992.

[98] Inoue T., and Raniecki B., Determination of thermal-hardening stress in steels by use of thermal-plasticity theory. Jal. Mech. Sol, 26: 187-212, 1978.

[99] Inoue T., and Wang Z., Coupling between stress, temperature, and metallic structures during processes involving phases transformations, Materials Science and Technology, 1: 845-50, 1985.

[100] Ikawa H., Shin S., Oshige H., and Mekuchi Y., Trans. J. W. S, 8:46, 1977.

[101] Iung, T., Roch, F., and Schmitt, J.H., Prediction of the mechanical properties of thermomechanical rolled steels, Int. Conf. On Thermomechanical Processing of steels and Others Materials, Chandra. T., Sakai, T., Ed., The Minerals, Metals and Mater. Soc., 2085-2091, 1997.

[102] Johnson W.A. and Mehl R.F., Reaction kinetics in process of nucleation and growth, Transactions of the A.I.M.E., 135: 416-45, 1939.

[103] Kachanov L.M., On creep rupture time, Izvestiya Akademii Nauk SSSR, Otdeleniya Tekhnicheskikh I Nauk, 8:26-31, 1958.

[104] Kachanov L.M., Time of the rupture process under creep conditions, Z.v, Akad, Nauk, SSR, nº 8:26, 1859.

[105] Kachanov L.M., Introduction to Continuum Damage Mechanics, Martinus, Nijhoff Publisher, BostonDordrecht, 1986.

[106] Kim K.T., Kwon Y.S., Strain hardening response of sintered porous iron tubes with various initial porosities under combined tension and torsion, J. Engrg. Mater. Technol. 114: 213-217, 1992.

[107] Knott J.F., Fundamentals of Fracture Mechanics, Butterworths, London, UK, 1988.

[108] Koistinen D.P. and Marburger R.E., A general equation prescribing extent of austenite-martensite transformation in pure Fe-C alloys and plain carbon steels, Acta Metallurgica, 7: 59-60, 1959.

[109] Kondo K., Ueda M., Ogawa K., Amaya H., Hirata H., and Takabe H., Alloy Design of Super 13 Cr Martensitic Stainless Steel (Development of Super 13 Cr Martensitic Stainless Steel for Line Pipe-1). In Supermartensitic Stainless Steels, 99: 11-18, Belgium, 1999.

[110] Koplik J., Needleman A., Void growth and coalescence in porous plastic solids, International Journal of Solids Structures, 24 (8): 835-853, 1988.

[111] Krajcinovic D., Constitutive equations for damaging materials, J. Appl. Mech., 50: 355-360, 1983.

[112] Lacombe P., Baroux B., and Beranger G., Stainless Steels, Les Editions de Physique Les Ulis, 1993.

[113] Lammer H., Tsamakis Ch., Discussion of coupled elastoplasticity and damage constitutive equations for small and finite deformations, International Journal of Plasticity, 16: 495-523, 2000.

[114] Leblond J.B. and Devaux J., A new kinetic model for anisothermal metallurgical transformations in steels including effect of austenite grain size. Acta Metal., 32(1): 137-146, 1984.

[115] Leblond J.B., Mottet G., Devaux J., Devaux J.C., Mathematical models of anisothermal phase transformations in steels, and predicted plastic behaviour, Mater. Sci. Tech, 1:815-822, 1985.

[116] Leblond J.B., Mottet G., and Devaux J.C., A theoretical and numerical approach to the plastic behavior of steels during phase transformations-I. Derivation of general relations, J. Mech. Phys. Solids 34: 395-409, 1986.

[117] Leblond J.B., Mottet G., and Devaux J.C., A theoretical and numerical approach to the plastic behavior of steels during phase transformations- Ⅱ. Study of classical plasticity for ideal-plastic phases, J. Mech. Phys. Solids 34:410-432, 1986.

[118] Leblond J.B., Devaux, J., and Devaux, J.C., Mathematical modelling of transformation plasticity in steels-I: Case of ideal-plastic phases, International Journal of Plasticity, 5:551-572, 1989.

[119] Leblond J.B., Devaux J., and Devaux J.C. Mathematical modelling of transformation plasticity in steels Ⅱ: Coupling with strain hardening phenomena, International Journal of Plasticity, 5: 573-591, 1989.

[120] Leblond J.B., Perrin, G., Devaux J., Bifurcation effects in ductile metals with nonlocal damage, ASME J. Appl. Mech., 61: 236-242, 1994.

[121] Leckie F. A., Hayhurst D. R., Constitutive equations for creep rupture, Acta. Metall, 25:1059-1107, 1977.

[122] Lemaitre J., A continuous damage mechanics model for ductile fracture, Journal of Engineering Material and Technology, 107: 83-89, 1985.

[123] Lemaitre J., Local approach to fracture. Engineering Fracture Mechanics, 25 (5-6): 523-537, 1986.

[124] Lemaitre J., Chaboche J.L., Mechanics of Solids Materials, Cambridge University Press, 1990.

[125] Lemaitre J., A Course on Damage Mechanics, Springer-Verlag, 1992.

[126] Liebaut C., Gautier E., and Simon A., Etude réologique d'un acier fe-0,2%c durant sa transformation de phase. Mémoire et étude scientifique Revue de métallurgie, 571-579, 1988.

[127] Lindgren L.-E., Finite element modelling and simulation of welding, Part 1. Increased complexity, J. Therm. Stresses, 24: 141-192, 2001.

[128] Lindgren L.-E., Finite element modelling and simulation of welding, Part 2, Improved material modelling, J. Therm. Stresses, 24:195-231, 2001.

[129] Lindgren L.-E., Finite element modelling and simulation of welding, Part 3. Efficiency and

integration, J. Therm. Stresses ,24: 305-334, 2001.

[130] Lindgren L.-E., Numerical modelling of welding, Computer Methods in Applied Mechanics and Engineering, 195(48-49): 6710-6736, 2006.

[131] Lu T.J., Chow C.L., On constitutive equations of inelastic solids with anisotropic damage, Theor. Appl. Fract. Mech., 14: 187-218, 1990.

[132] Magee C.L., Nucleation of martensite, Phases transformations, ASM, Metals Park, 1970.

[133] Magee C.L. Transformation kinetics, microplasticity and ageing of martensite in Fe-31-Ni, Ph. D. Thesis Carnegie Mellon University, Pittsburg, 1966.

[134] Mariage J.-F., Simulation numérique de l'endommagement ductile en formage de pièces massives, université de technologie Troyes, Thèse, 2003.

[135] Martinez M., Jonction 16MND5-INCONEL 690-316LN par soudage-diffusion, PhD Thesis, ENSMP, Paris, France, 1999.

[136] Maruyama K., Sawada K., and Koike J.I., Strengthening mechanisms of creep resistant tempered martensitic steel. ISIJ international. 41(6): 641-653, 2001.

[137] Materials Science and Metallurgy, 4th ed., Pollack, Prentice-Hall, 1988.

[138] McClintock F.A., A criterion for ductile fracture by the growth of holes, Journal of Applied Mechanics, 35: 363-371, 1968.

[139] Murakawa H., Prediction of welding deformation and residual stress by elastic FEM based on inherent strain, Trans. Soc, Naval Architects of Japan, 180: 739-751, 1996

[140] Murakami S., Ohno N., A continuum theory of creep and creep damage, In: Ponter, A.R.S., Hayhurst, D.R. (Eds.), Creep in Structures, Springer Verlag, Berlin, 422-443, 1981.

[141] Murakami S., Mechanical modeling of material damage, J. Appl. Mech., 55: 280-286, 1988.

[142] Needleman A., Tvergaard V., An analysis of ductile rupture in notched bars, Journal of Mechanics and Physics of Solids 32: 461, 1984.

[143] Noyan I.C., and Cohen J.B., Residual stress—measurement by diffraction and interpretation, in: B. Ilschner, N.J. Grant (Eds.), Materials Research and Engineering Series, Springer-Verlag, NY, 1987.

[144] Oberkampf W.L., and Trucano T.G., Verification and validation in computational fluid dynamics, Prog. Aerosp. Sci. 38 (3): 209-272, 2002.

[145] Oddy A.S., Goldak J., McDill M.J., Transformation plasticity and residual stresses in

single-pass repair welds, J. Pressure Vessel Tech, 11: 33-38, 1992.

[146] Olson G.B., and Cohen M., Kinetics of strain-induced martensitic nucleation, Metall. Trans. 6A, 791-795, 1975.

[147] Olson G.B., and Cohen M., Stress-assisted isothermal martensitic transformation: application to TRIP steels, Metall. Trans. 13A: 1907-1913, 1982.

[148] Ortin J., Thermodynamics and kinetics of phase transitions: an Introduction, In: Mechanics of solids with phase changes, Berveiller M. and Fischer F. D.(eds), 1-52, 1997.

[149] Pardoen T., Hutchinson J.W., An extended model for void growth and coalescence, J. Mech. Phys. Solids, 48: 2467-2512, 2000.

[150] Patel J.R. and Cohen M., Criterion for the action of applied stress in the martensite transformation. Acta Mater., 1: 531-538, 1953.

[151] Peters W.H., and Ransom W.F., Digital imaging techniques in experimental stress analysis, Opt. Eng., 21(3): 427-431, 1892.

[152] Petch N., Fracture, John Wiley & son, New York, USA, 1959.

[153] Petit-Grostabussiat. S., Conséquences mécaniques des transformations structurales dans les alliages ferreux, Thèse de Doctorat, INSA de Lyon, 2000.

[154] Pirondi A., Bonora N., Modeling ductile damage under fully reversed cycling, Computational Materials Science, 26: 129-141, 2003.

[155] Porter, L.F., and Rosenthal, P.C, Effect of applied tensile stress on phase transformation in steel, Acta metall, 7: 504-514, 1959.

[156] Priestner R., Ibraheem A.K., Mater. Sci. Technol. 16: 1267, 2000.

[157] Qiao G. Y., Zhang K.Q., and Xiao F. R., Effect of cooling rate on martensite microstructure of Fv520(B) steel, China Journal of Heat Treatment of Metals, No.2P: 31-32, 2000.

[158] Rabotnov Y.N., Creep problems in structural members, North-Holland, 1969.

[159] Ramberg W., Osgood W.R., Determination of stress-strain curves by three parameters, Technical note no. 503, National Advisory Committee on Aeronautics (NACA), 1941.

[160] Rice J.R., Tracey D.M., On ductile enlargement of voids in triaxial stress fields, Journal of Mechanics and Physics of Solids, 17: 210-217, 1969.

[161] Richmond O., and Smelser R.E., Alcoa Technical Center Memorandum, In: Hom, C.L. and McMeeking, R.M., Editors, 1985. Void growth in elastic-plastic materials, J. Appl. Mech.,

56: 309-317, 1989.

[162] Ronda J., Oliver G.J., Comparison of applicability of various thermo-viscoplastic constitutive models in modeling of welding, Comput. Methods Appl. Mech. Eng, 153: 195-221, 1998.

[163] Rousselier G., Three dimensionnal constitutive relations and ductile fracture, IUTAM Symp. On three dimensionnal constitutive relations and fraction, 197-226, Dourdan, 1981.

[164] Rousselier G., Devaux J. C., Mottet G. and Devesa G., A methodology for dutile fracture analysis based on damage mechanics: an illustration of a local approach of fracture, Nonlinear fracture mechanics: volume Ⅱ - Elastic plastic fracture, ASTM STP 995: 322-354, 1989.

[165] Rousselier G., in: J. Lemaitre (Ed.), Handbook of materials behavior models, Academic press, San Diego, CA, vol. 2: 436-445, 2001.

[166] Rybicki E. et al., A finite element model for residual stresses in girth-butt welded pipes, in: Numerical Modeling of Manufacturing Processes, ASME Winter Annual Meeting, 1977.

[167] Saanouni K., Cherouat A., Mariage J.F., Nesnas K., On the numerical simulation of 3D forging of some complex parts considering the damage occurrence, in: Proceedings of the Esaform, 2: 593-596, 2001.

[168] Saanouni K., and Chaboche J.L., Computational Damage Mechanics. Application to Metal Forming, In: de Borst, R. and Mang, H.A. (eds), Numerical and Computational Methods, Chapter 7, Vol. 3, In: Milne, I., Richie, R.O. and Karihabo, B., Comprehensive Structural Integrity, Oxford, UK, 321-376. 2003.

[169] Saanouni K., On the numerical prediction of the ductile fracture in metal forming, Eng Fract Mech, Article in Press, 2007.

[170] Schacht T., Untermann N., Steck, E., The influence of crystallographic orientation on the deformation behavior of single crystals containing microvoids, International Journal of Plasticity, 19:1605-1626, 2003.

[171] Simon A., Denis, S., and Gautier E., Effet des sollicitations thermo-mécaniques sur les transformations de phases dans l'état solide. Aspects métallurgique et mécanique, Journal de Physique Ⅳ, Colloque C3 supplément au journal de physique Ⅲ, 4: 199-213, 1994.

[172] Sjostrom S., Interactions and constitutive models for calculating quench stresses in steel. Mat. Sci. Tech., 1: 823-829, 1985.

[173] Staub C. and Boyer J.C., A ductile growth model for elasto-plastic material, Journal of Materials Processing Technology, 77(1-3): 9-16, 1998.

[174] Steinmann P., Carol I., A framework for geometrically nonlinear continuum damage mechanics, Int. J. Eng. Sci., 36: 1793-1814, 1998.

[175] Stringfellow R.G., Mechanics of strain-induced transformation toughening in metastable austenitic steels, Phdthesis, MIT, Cambridge, 1990.

[176] Stringfellow R.G., Parks D.M., Olson G.B., A constitutive model for transformation plasticity accompanying strain-induced martensitic transformation in metastable austenitic steels, Acta Metall. Mater. 40: 1703-1716, 1992.

[177] Stuwe H.P., Interaction of Stresses and Strains with Phase Changes in Metals: Physical Aspects. In Mechanics of solids with phase changes, Berveiller M. and Fischer F. D.(eds), 53-68, 1997.

[178] Sun Y., Wang D., A lower bound approach to the yield loci of porous materials, Acta Mech. Sinica 5:(3): 237-243, 1989.

[179] Suresh S., and Giannakopoulos A.E., Acta Mater. 46: 5755, 1998.

[180] Tai H.W., Yang B.X., A new microvoid-damage model for ductile fracture, Engineering Fracture Mechanics, 25 (3): 377-384, 1986.

[181] Tai H.W., Plastic damage and ductile fracture in mild steels, Engineering Fracture Mechanics 36 (4): 853880, 1990.

[182] Taleb L., and Sidoroff. F., Micromechanical modelling of greenwood-johnson mechanism in transformation induced plasticity, Int. Jal. Plas., 19: 1821-1842(22), 2003.

[183] Taylor, GI, Plastic strains in metals, Journal of the Institute of Metals, 62: 307-24, 1938.

[184] Tekriwal P., Mazumder J., Transient and residual thermal strain-stress analysis of GMAW, J. Eng. Mater. Technol, 113: 336-343, 1991.

[185] Touchal S., Morestin F., Brunet M., Use of damage model and method by correlation of digital images for the necking detection, 14th international Conference on SMIRT, Lyon, France, 1997.

[186] Trillat M., Pastor J., Limit analysis and Gurson's model, European Journal of Mechanics-A/Solids, 24(5): 800-819, 2005.

[187] Trillat M., Pastor J., and Francescato P., Yield criterion for porous media with spherical

voids, Mechanics research communications, 33(3): 320-328, 2006.

[188] Tvergaard V., Needleman A., Analysis of the cup-cone fracture in a round tensile bar, Acta Metall, 32 (1): 157-169, 1984.

[189] Tvergaard V., Mechanical modelling of ductile fracture, Mechanica, 26: S11-6, 1991.

[190] Tvergaard V., Niordoson C., Non local plasticity effects on interaction of different size voids. International Journal of Plasticity, 20: 107-120, 2004.

[191] Ueda Y., and Yamakawa T., Analysis of thermal elastic-plastic stress and strain during welding by finite element method, Trans. Jpn. Weld. Res. Inst.2: 90-100, 1971.

[192] Ueda Y., A prediction method of welding residual stress using source of residual stress (inherent strain), Trans. JWRI, 18(1): 135-14, 1989.

[193] Valance S., Coret M., Combescure A., Strain simulation of steel during a heating-cooling cycle including solid-solid phase change, European Journal of Mechanics-A/Solids, 26(3): 460-473, 2007.

[194] Verpeaux P., Charras T., and Millard A., CASTEM 2000 : une approche moderne du calcul des structures. In J.-M. Fouet, P. Ladevèze, and R. Ohayon, editors, Calcul des Structures et Intelligence Artificielle, 2: 261-271, 1988.

[195] Videau J.C., Cailletaud G., and Pineau A., Modélisations des effets mécaniques des transformations de phase, Jal. de Ph., 3: 227-232, 1994.

[196] Videau, J.C., Cailletaud, G., and Pineau, A. Experimental study of the transformation induced plasticity in a Cr-Ni-Mo-Al-Ti steel, Journal de Physique Ⅳ, colloque C1, Supplément au J de Physique Ⅲ, 6:465474, 1995.

[197] Vincent Gaggard, Experimental study and modelling of high temperature creep flow and damage behaviour of 9Cr1Mo-Nbv steel weldments, Ecole des Mines de Paris, These, 2004.

[198] Vincent Y., Simulation numerique des consequenes metallugriques et mecaniques induites pqr une operation de soudage: acier 16MND5, Thèse de Doctorat, INSA de Lyon, 2002.

[199] Vincent Y., Jullien J.F., Gilles P., Thermo-mechanical consequences of phase transformations in the heataffected zone using a cyclic uniaxial test, International Journal of Solids and Structures, 42(14): 40774098, 2005.

[200] Voyidajis G., Abu Al-Arub R.K., and Palazzotto A.N., Thermodynamic framework for coupling of nonlocal viscoplasticity and non-local anisotropic viscodamage for dynamic

localization problems using gradient theory, International Journal of Plasticity, 20: 981-1038, 2004.

[201] Waeckel F., Une loi de comportement thermo-métallurgique des aciers pour le calcul mécanique des structures (A thermo-metallurgical constitutive law of steels for structural mechanics), Thèse, ENSAM, 1994.

[202] Waeckel F., Dupas P., and Andirieu S., A thermo-metallurgical model for steel cooling behavioue : proposition, validation and comparison with the sysweld's model, J. Ph., C1, Ⅲ(6): 255-263, 1996.

[203] Wang J.H., An FEM model of buckling distortion during welding of thin plate, J. of Shanghai Jiaotong University, E-4(2), 81-96, 1997.

[204] Wattrisse B., Chrysochoos A., Muracciole J.-M. and Némoz-Gaillard M., Kinematic manifestations of localisation phenomena in steels by digital image correlation, European Journal of Mechanics A/Solids, 20(2): 189-211, 2001.

[205] Webster G.A., and Ezeilo A.N., Int. J. Fatigue 23, 2001.

[206] Wert C., Interference of growing spherical precipitate particles, J. Appl. Phy., 21: 5-8, 1950.

[207] Wu T., Coret M., Combescure A., Phase transformation and damage elastoplastic multiphase model for welding simulation, Proceedings of the 16th European Conference of Fracture, Alexandroupolis, Greece, 2006, In: Fracture of Nano and Engineering Materials and structures, Gdoutos E.E.(Ed.), Springer, 2006.

[208] Wu T., Coret M., Combescure A., A Mesoscopic approach to simulate damage induced by welding of 15-5PH, the 8th International Seminar on the Numerical Analysis of Weldability, Graz-Seggau, Austria, 2006.

[209] Wu T., Gilles P., Coret M., Combescure A., Welding damage numerical model of 15-5PH stainless steel, Conference of MATERIAUX2006, Dijon, France, 2006.

[210] Xia L., Shih C.F., Ductile Crack Growth-I, A numerical study using computational cells with microstructurally-based length scales, Journal of the Mechanics and Physics of Solids, 43 (2): 233-259, 1995.

[211] Zhou S. F., et al, Microstructure and mechanical properties in simulated HAZ of 0Cr13Ni5Mo martensitic stainless steel, Transactions of the china welding institution, 25(5): 63-66, 2004.

Appendix A Experimental Devices

A.1 Strain and stress measurements

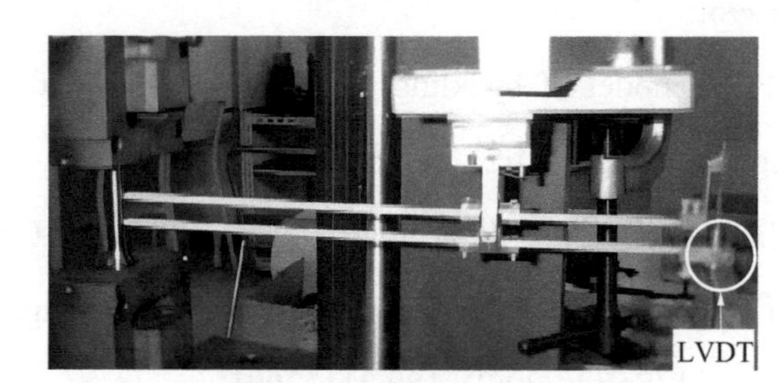

a) Figure

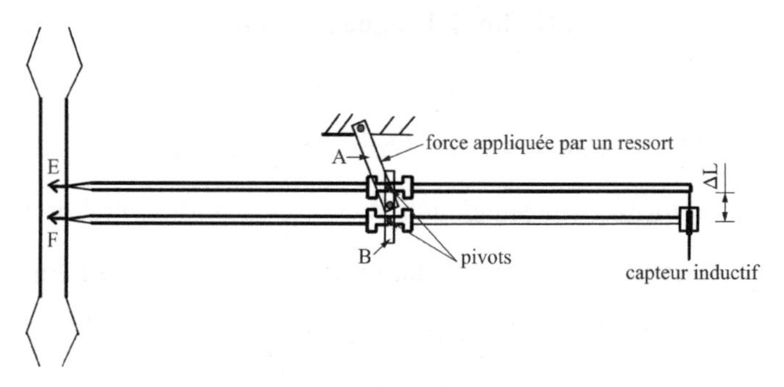

b) Diagram

Figure A.1 Extensometer

The linear variable differential transformer (LVDT) is a type of electrical transformer used for measuring linear displacement. The transformer has three solenoidal coils placed end-to-end around a tube. The centre coil is the primary, and the two outer coils are the secondaries. A cylindrical ferromagnetic core, attached to the object whose position is to be measured, slides along the axis of the tube.

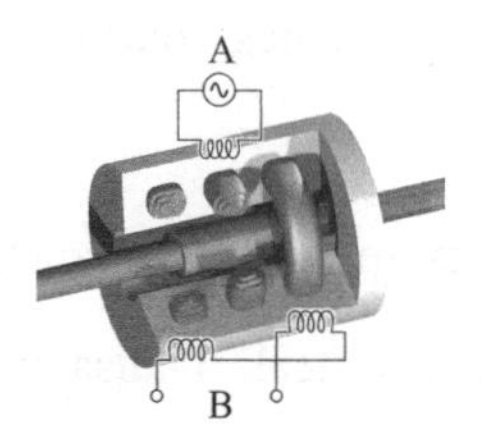

Figure A.2 LVDT

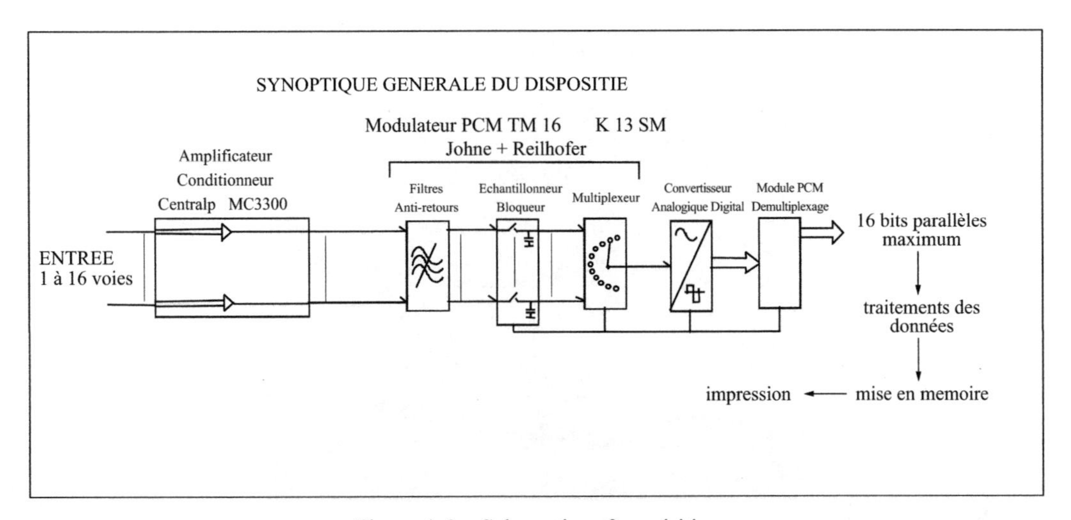

Figure A.3 Schematics of acquisition

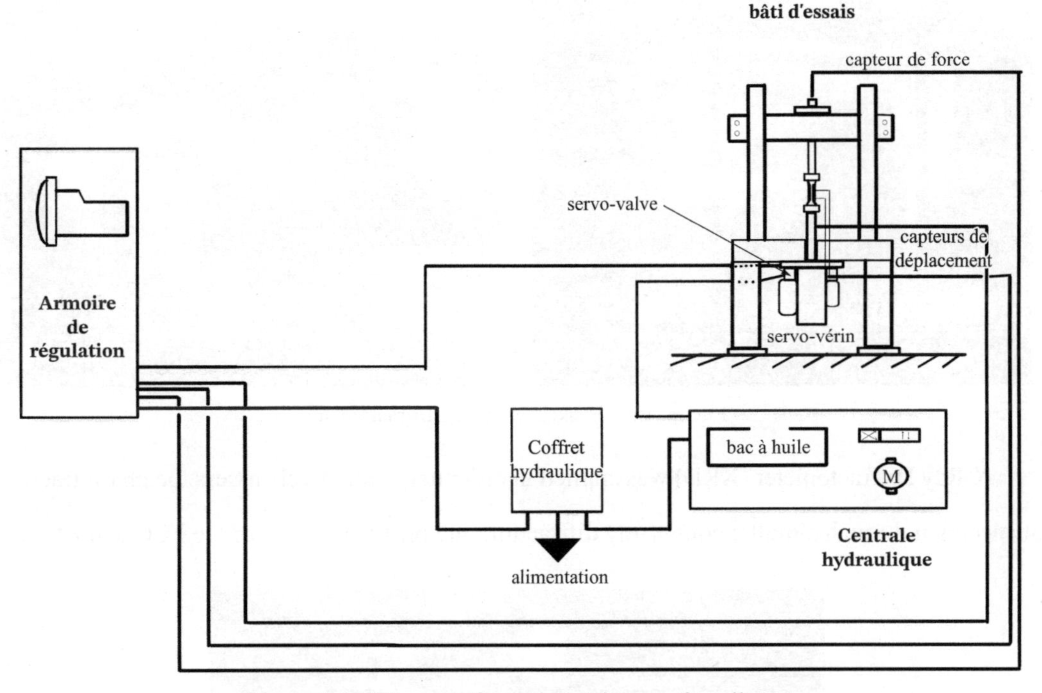

Figure A.4 Diagram of experimental setup of tensile test

A.2 Microscopic equipments

The optical microscopy was used to observer topologic image of microstructures (grain size, grain boundary) and crack region of stretched round bar. The specimen was prepared by standard methods involving mechanical grinding (ground using emery papers of various grinds from 200 to

2500 grit), polishing (silicon grit from 6μm to 1μm) and vibration-polished.

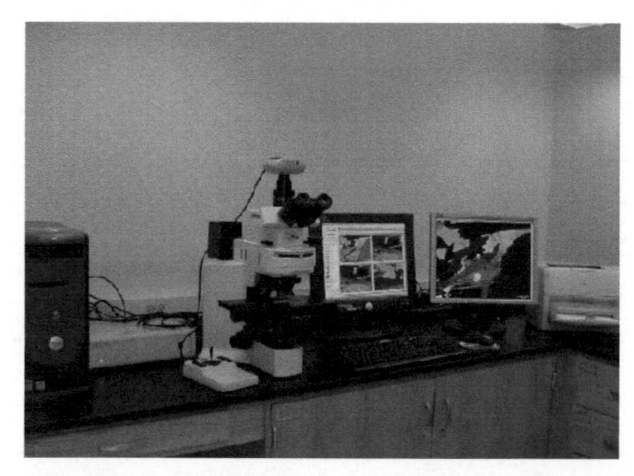

Figure A.5 Optical microscopy (ZEISS AX10PHOT)

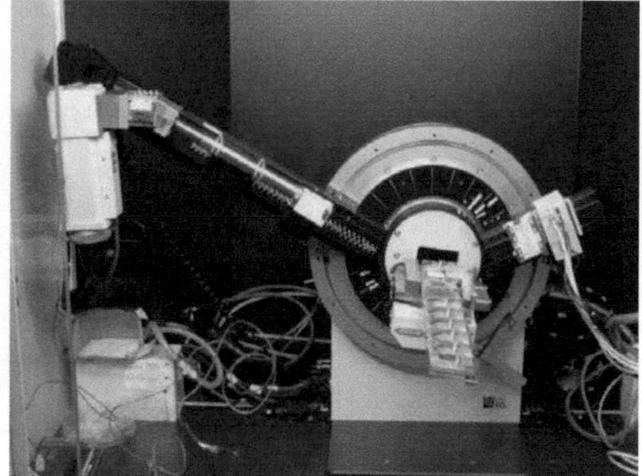

Figure A.6 Bruker D8 X-ray diffractometer

X-Ray Diffractometer (XRD) was applied in order to quantitatively assess the phase fractions of specimen through simultaneous X-ray diffraction line profile analysis of the FCC and BCC.

Figure A.7 TEM JEOL 2010 FEG

Transmission Electron Microscope (TEM) was implemented to study the microstructure of the 15-5PH in the as-received condition, and analyze the structure of the precipitates. The characteristics of the TEM JEOL 2010 FEG is as following: Accelerating voltage 200 kV , field emission source, point resolution 2.0 Å, useful limit of information ~1.2 Å, maximum specimen tilt capacity +/−30 degrees, absolute minimum probe size ~0.4 nm, EDS spatial resolution ~1 nm, minimum energy spread ~0.65 eV.

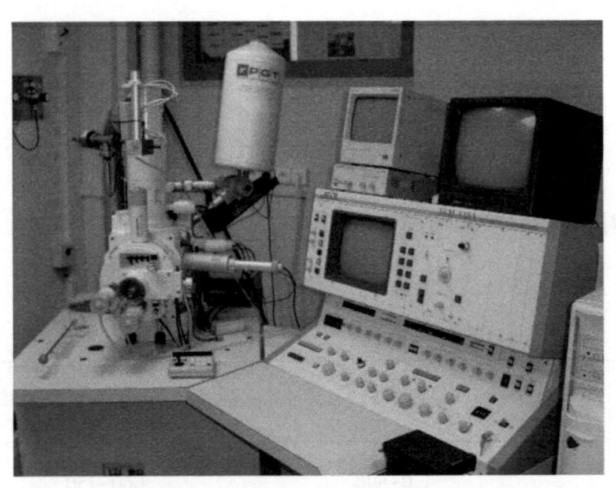

Figure A.8 SEM JEOL 840 A LGS

Scanning Electron Microscope (SEM) was carried out for purpose of observing the microscopic characteristics including microvoids and microcracks. The specimen was prepared with the same procedure of the optical electron microscopy observation: mechanical grinding (ground using emery papers of various grinds from 200 to 2500 grit), polishing (silicon grit from 6μm to 1μm) and vibration-polished for observing surface.

Appendix B Experimental Results of 15-5PH in Details

B.1 Test results of round bar

B.1.1 Material properties

Expansion coefficients — Table B.1

	Martensite		Austenite
Temperature (°C)	0~600	660~750	200~1000
ALFA (10^{-6}/°C]	13.5	16.8	21.6

Young's modulus — Table B.2

Temperature (°C)		20	200	600	700	850
E (GPa)	Martensite	199	189	121	78	—
	Austenite	—	180	120	—	38

Yield strength and ultimate strength — Table B.3

Temperature (°C)		20	200	600	700	850
Martensite	Nonlinear SIG_Y (MPa)	861	782	322	108	—
	0.2% SIG_Y (MPa)	1028	901	455	159	—
	Ultimate SIG (MPa)	1164	1026	527	290	—
Austenite	Nonlinear SIG_Y (MPa)	—	115	98	—	77
	0.2% SIG_Y (MPa)	—	169	106	—	115
	Ultimate SIG (MPa)	—	521	388	—	177

B.1.2　Force vs. displacement

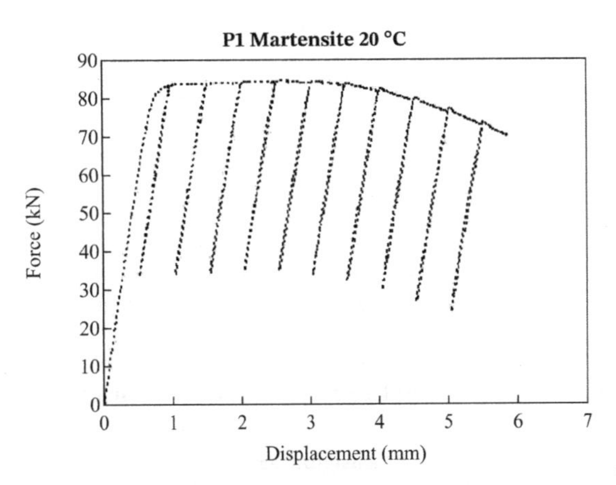

Figure B.1　Force vs. displacement of P1 at 20 °C, martensitic state

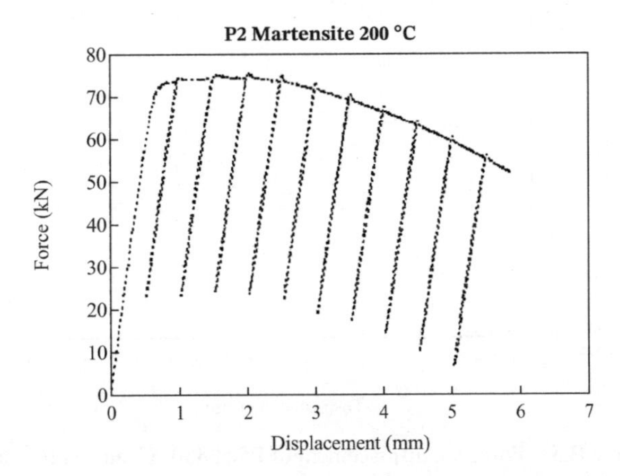

Figure B.2　Force vs. displacement of P2 at 200 °C, martensitic state

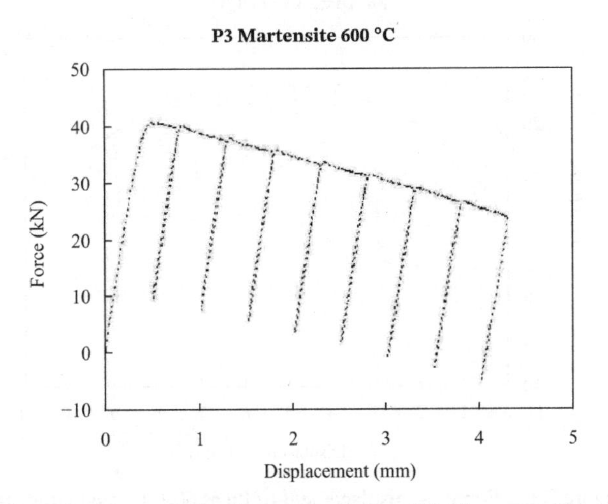

Figure B.3　Force vs. displacement of P3 at 600 °C, martensitic state

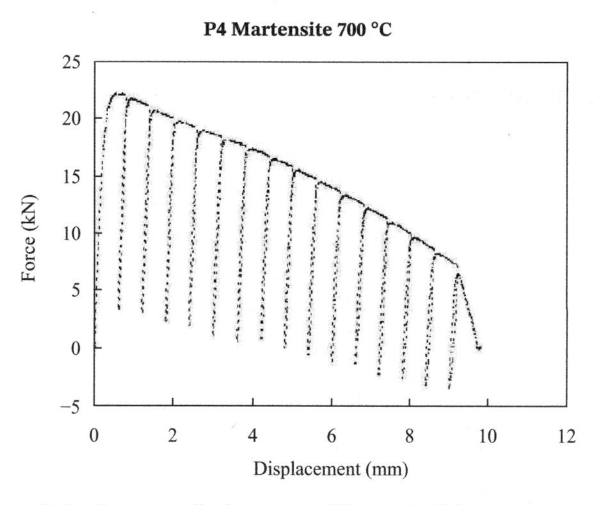

Figure B.4 Force vs. displacement of P4 at 700 °C, martensitic state

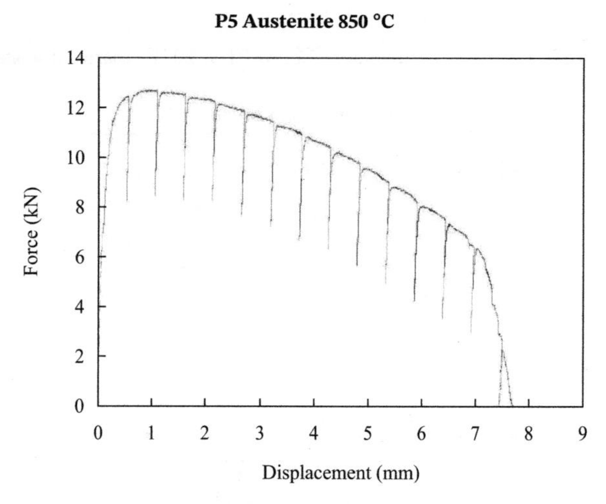

Figure B.5 Force vs. displacement of P5 at 850 °C, austenitic state

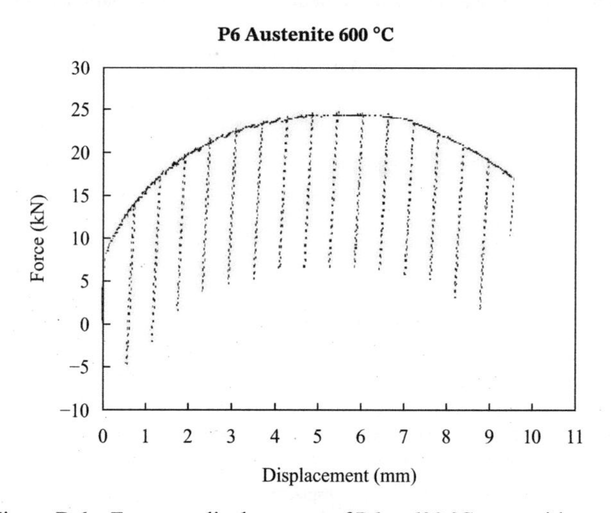

Figure B.6 Force vs. displacement of P6 at 600 °C, austenitic state

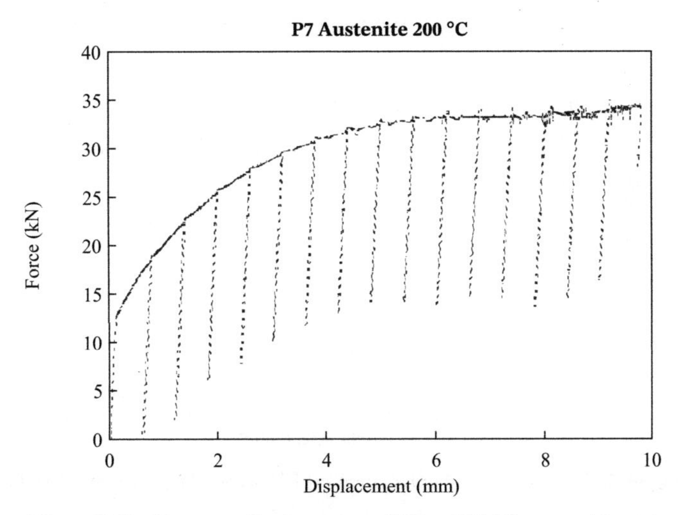

Figure B.7 Force vs. displacement of P7 at 200 °C, austenitic state

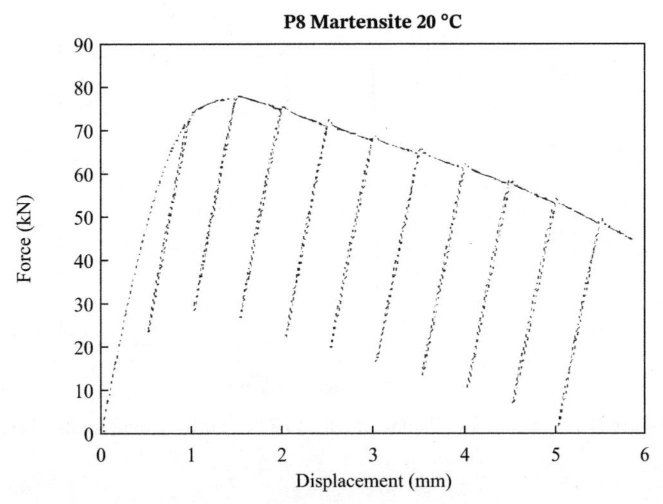

Figure B.8 Force vs. displacement of P8 at 20 °C, martensitic state

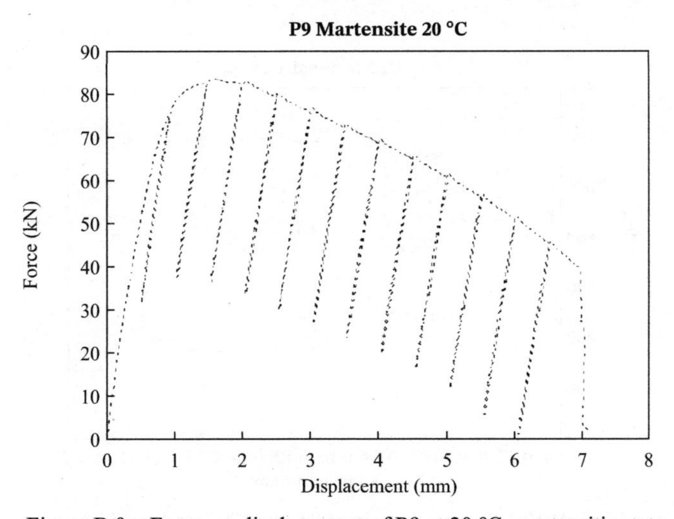

Figure B.9 Force vs. displacement of P9 at 20 °C, martensitic state

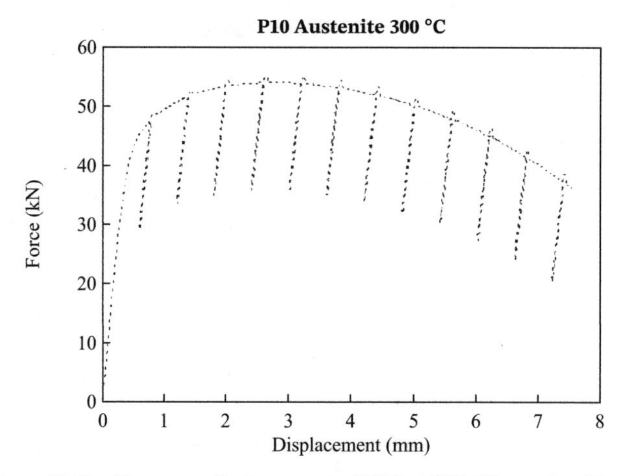

Figure B.10 Force vs. displacement of P10 at 300 °C, martensitic state

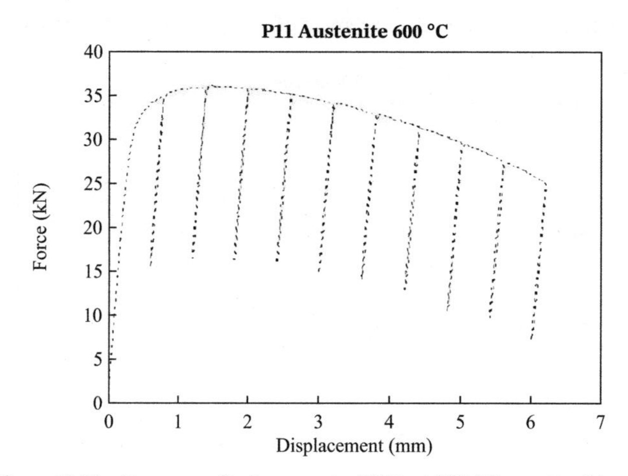

Figure B.11 Force vs. displacement of P11 at 600 °C, martensitic state

B.1.3 Stress vs. strain

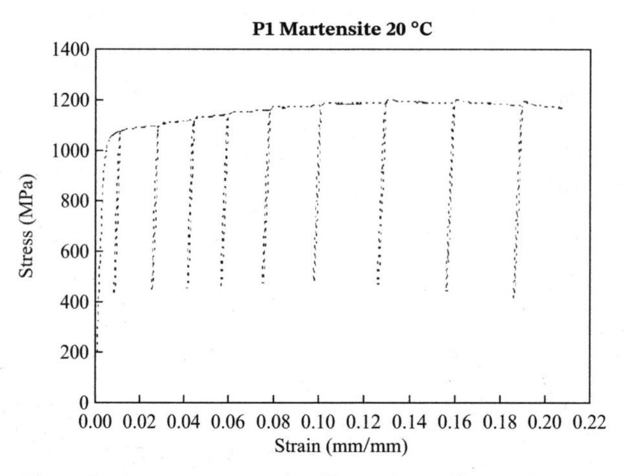

Figure B.12 Stress vs. strain of P1 at 20 °C, martensitic state

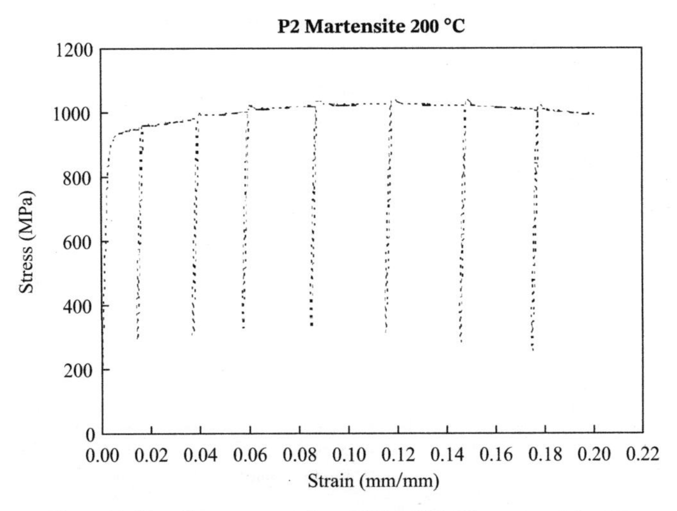

Figure B.13 Stress vs. strain of P2 at 200 °C, martensitic state

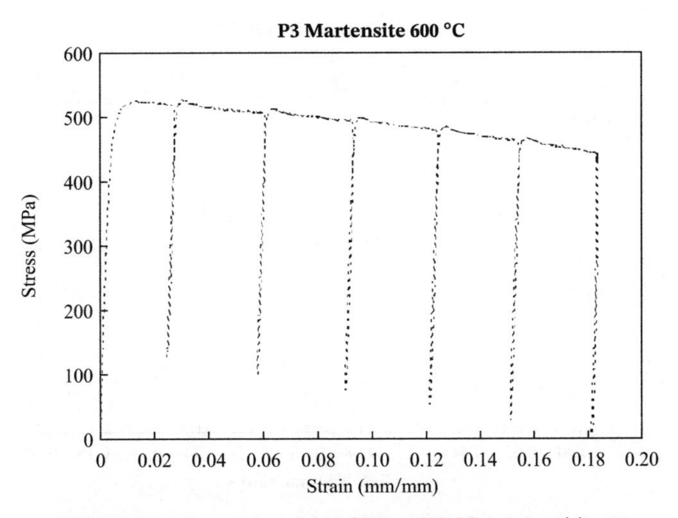

Figure B.14 Stress vs. strain of P3 at 600 °C, martensitic state

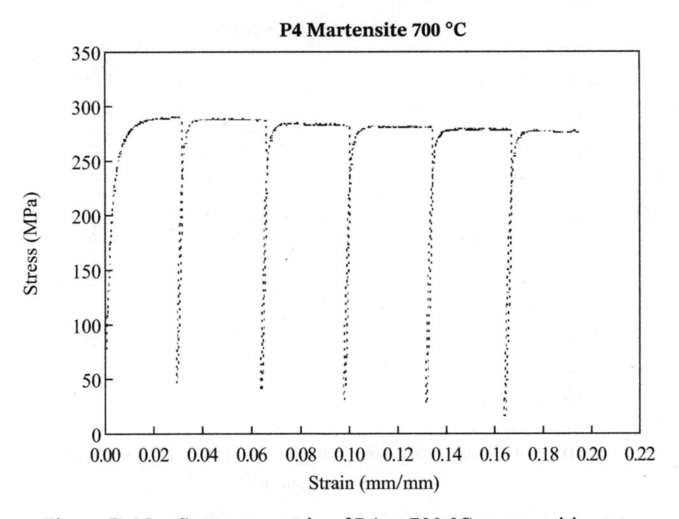

Figure B.15 Stress vs. strain of P4 at 700 °C, martensitic state

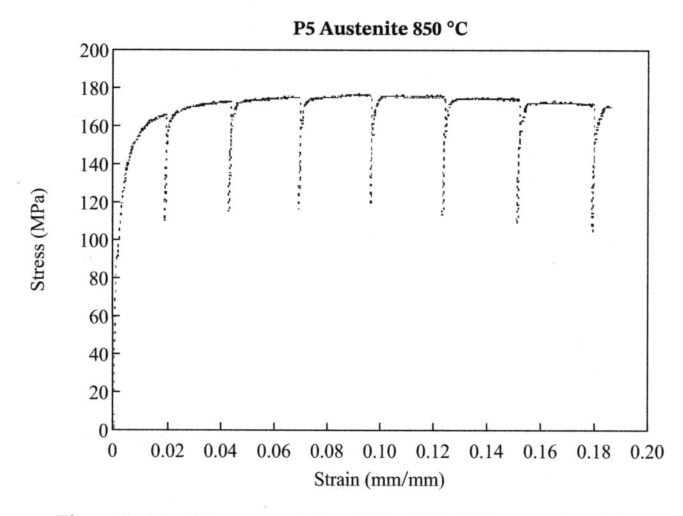

Figure B.16 Stress vs. strain of P5 at 850 °C, austenitic state

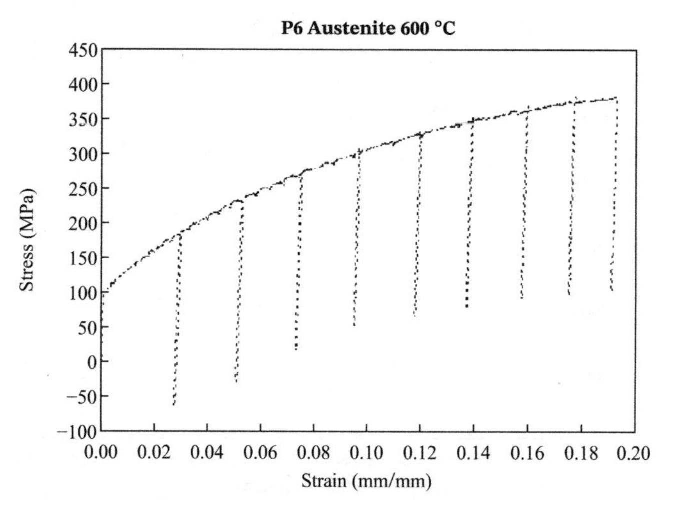

Figure B.17 Stress vs. strain of P6 at 600 °C, austenitic state

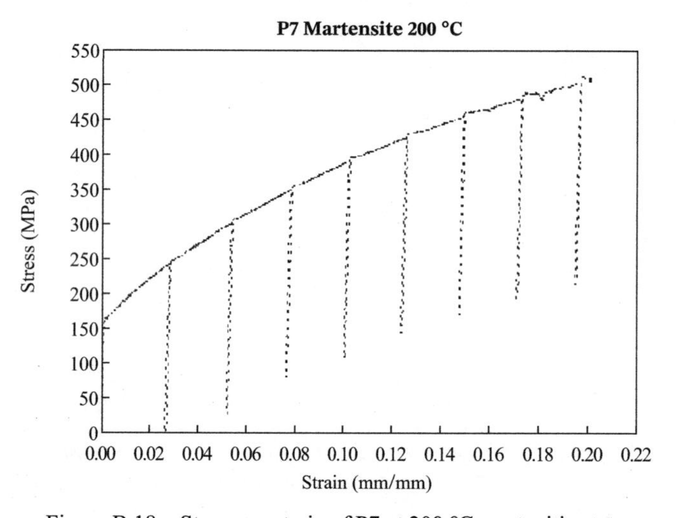

Figure B.18 Stress vs. strain of P7 at 200 °C, austenitic state

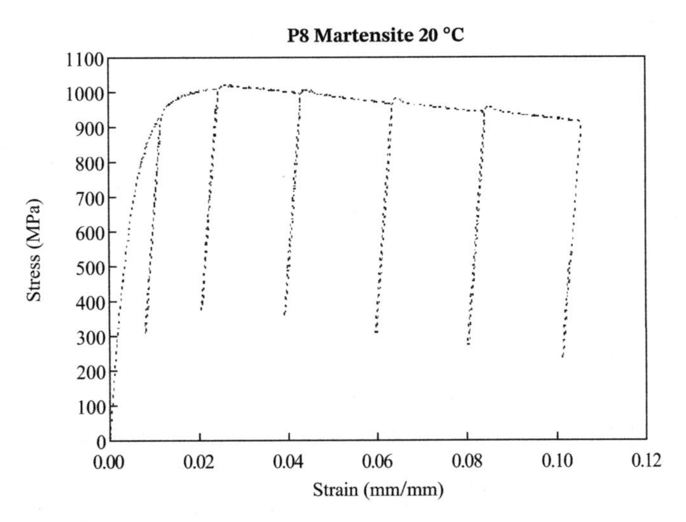

Figure B.19 Stress vs. strain of P8 at 20 °C, martensitic state

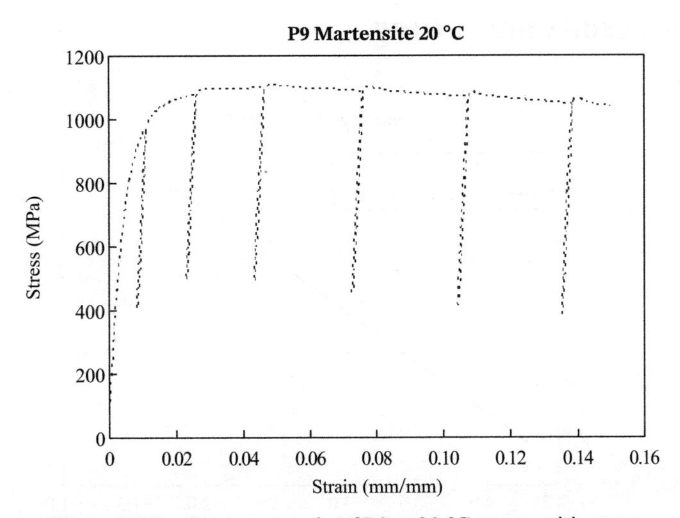

Figure B.20 Stress vs. strain of P9 at 20 °C, martensitic state

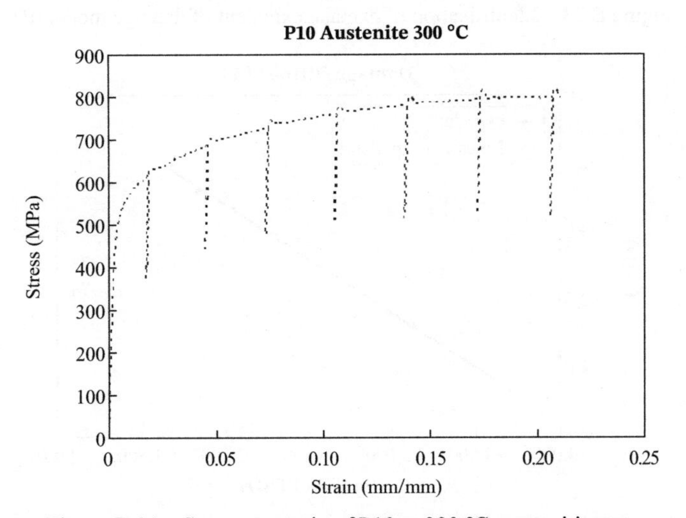

Figure B.21 Stress vs. strain of P10 at 300 °C, austenitic state

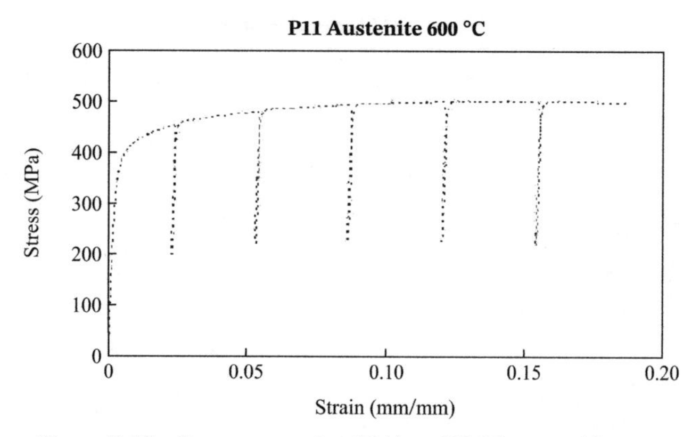

Figure B.22 Stress vs. strain of P11 at 600 ℃, austenitic state

B.2 Damage results and fitting

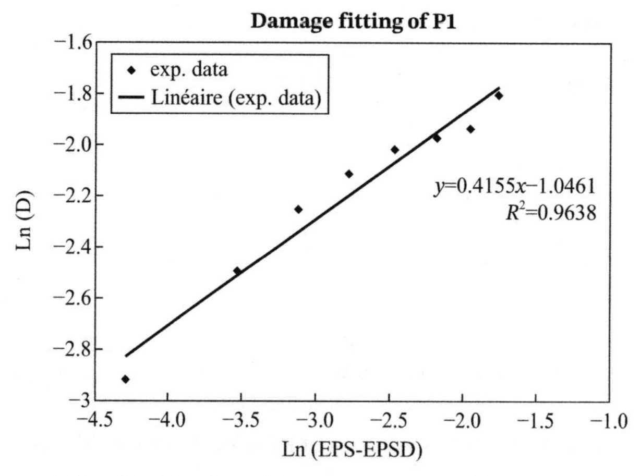

Figure B.23 Identification of damage exponent of damage model (P1)

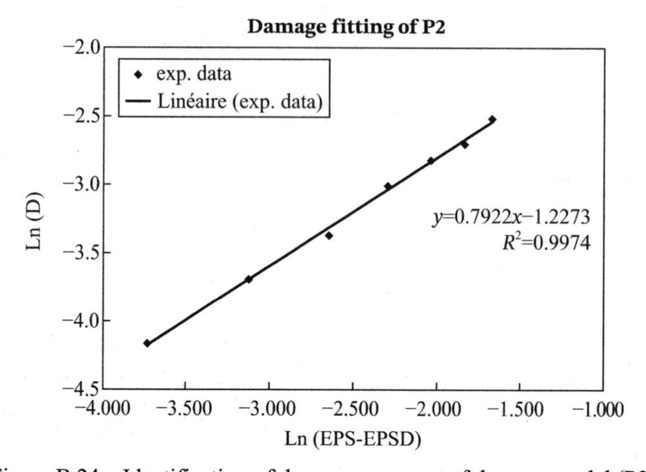

Figure B.24 Identification of damage exponent of damage model (P2)

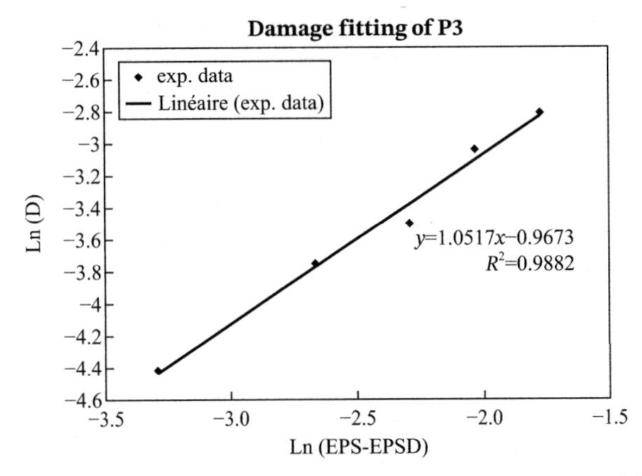

Figure B.25 Identification of damage exponent of damage model (P3)

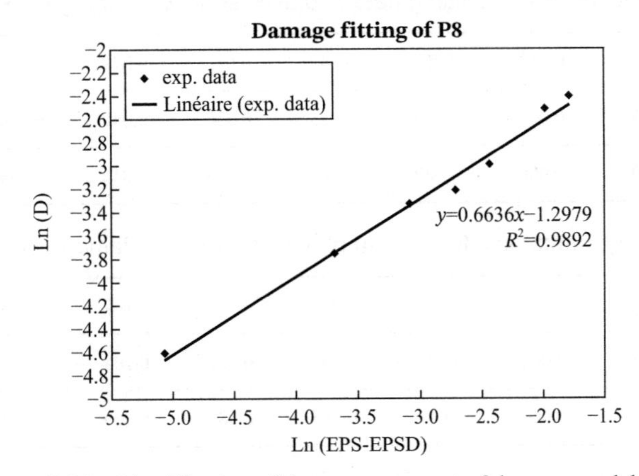

Figure B.26 Identification of damage exponent of damage model (P8)

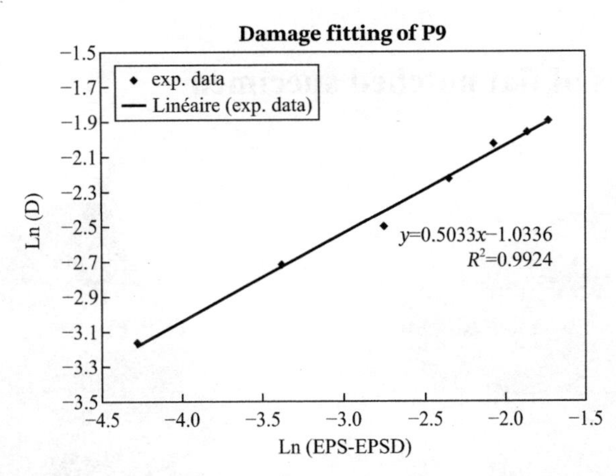

Figure B.27 Identification of damage exponent of damage model (P9)

Experimental and fitting data of plastic strain vs. damage of P1 Table B.4

Plastic strain	0.000	0.007	0.024	0.039	0.055	0.073	0.095	0.123	0.153	0.183	0.210
Damage (exp.)	0.000	0.000	0.054	0.082	0.105	0.121	0.132	0.139	0.144	0.164	0.180
Damage (fitting)	0.000	0.000	0.060	0.082	0.098	0.113	0.128	0.144	0.159	0.172	0.183

Experimental and fitting data of plastic strain vs. damage of P2 Table B.5

Plastic strain	0.000	0.014	0.036	0.056	0.083	0.113	0.143	0.172	0.200
Damage (exp.)	0.000	0.000	0.016	0.025	0.034	0.049	0.059	0.067	0.080
Damage (fitting)	0.000	0.000	0.015	0.025	0.036	0.048	0.058	0.068	0.078

Experimental and fitting data of plastic strain vs. damage of P3 Table B.6

Plastic strain	0.000	0.024	0.057	0.090	0.121	0.151	0.190
Damage (exp.)	0.000	0.000	0.012	0.024	0.030	0.048	0.060
Damage (fitting)	0.000	0.000	0.012	0.023	0.034	0.045	0.059

Experimental and fitting data of plastic strain vs. damage of P8 Table B.7

Plastic strain	0.000	0.012	0.018	0.037	0.058	0.079	0.100	0.150	0.180
Damage (exp.)	0.000	0.000	0.010	0.023	0.036	0.040	0.050	0.080	0.090
Damage (fitting)	0.000	0.000	0.009	0.024	0.035	0.045	0.054	0.073	0.084

Experimental and fitting data of plastic strain vs. damage of P9 Table B.8

Plastic strain	0.000	0.011	0.026	0.046	0.076	0.107	0.138	0.168	0.189
Damage (exp.)	0.000	0.000	0.042	0.066	0.082	0.108	0.131	0.140	0.150
Damage (fitting)	0.000	0.000	0.041	0.065	0.089	0.109	0.125	0.140	0.149

B.3 Test results of flat notched specimen

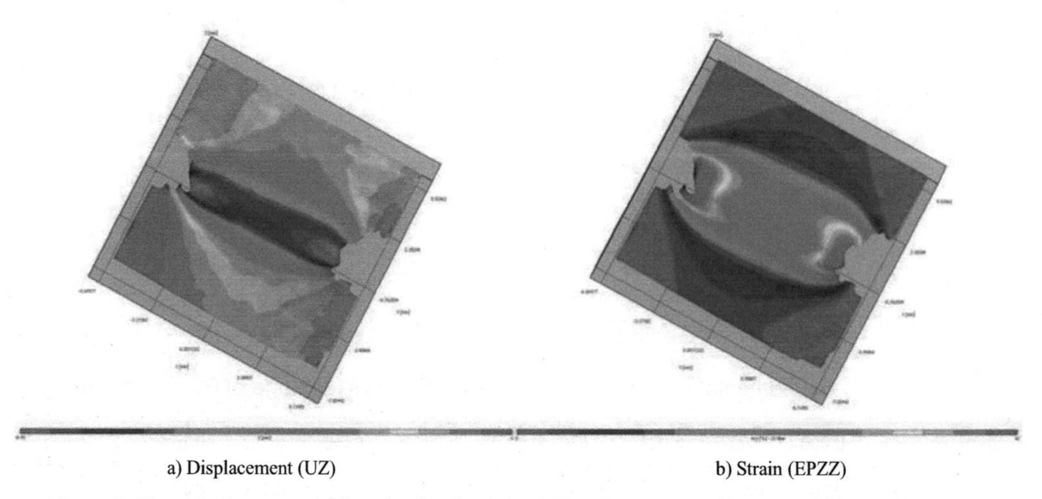

a) Displacement (UZ) b) Strain (EPZZ)

Figure B.28 Displacement and strain distribution of F2A (Case A, 200°C, martensite) when loading displacement is equal to 1.5mm (d = 1.5mm) by using digital image correlation

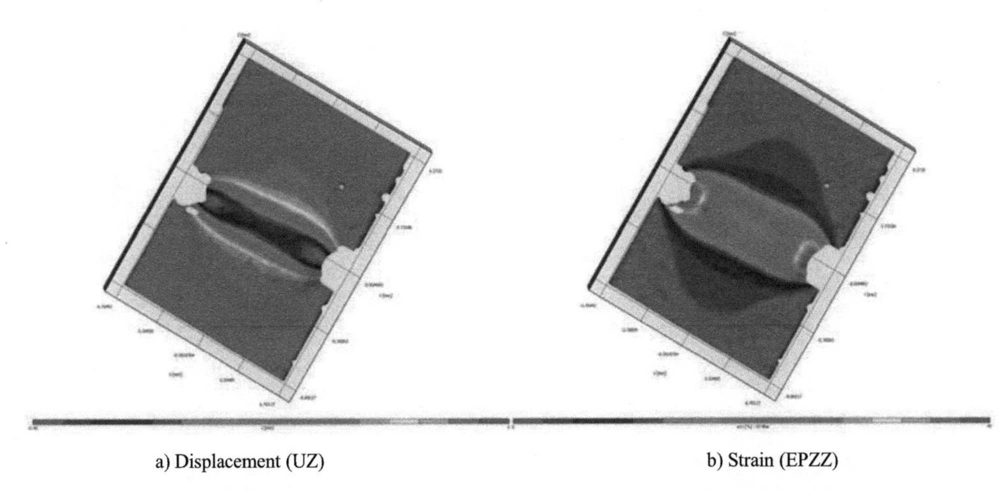

Figure B.29 Displacement and strain distribution of F3A (Case A, 20℃, mlartensite) when loading displacement is equal to 1.5mm (d = 1.5mm) by using digital image correlation

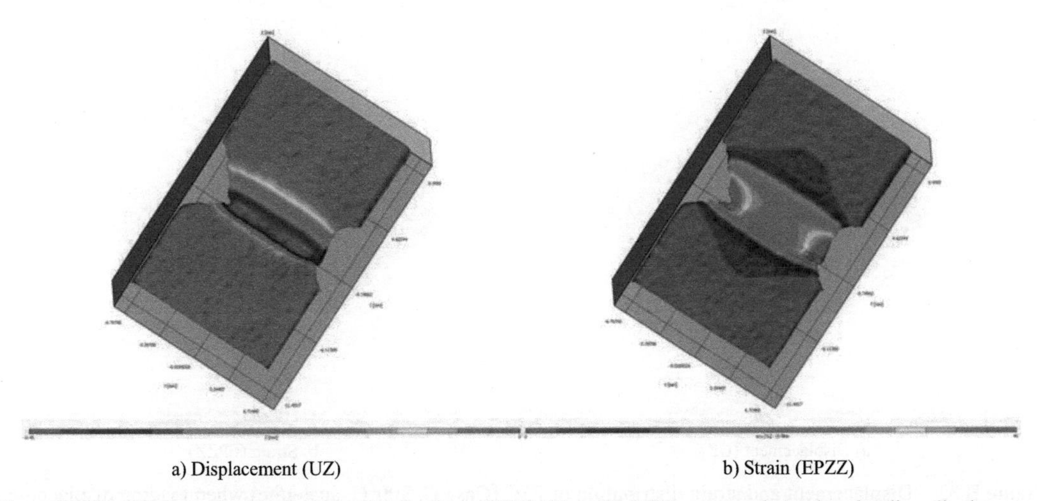

Figure B.30 Displacement and strain distribution of F3B (Case B, 20℃, martensite) when loading displacement is equal to 1.5mm (d = 1.5mm) by using digital image correlation

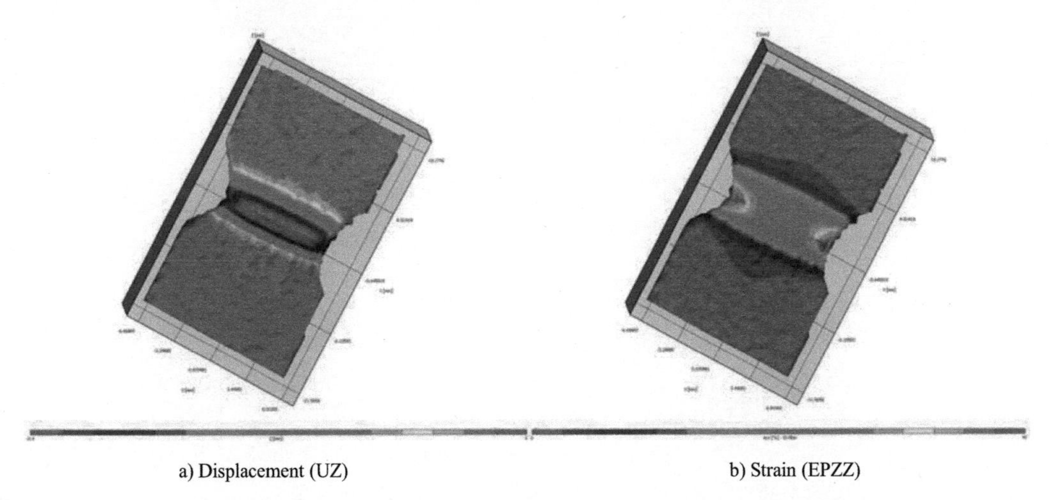

Figure B.31 Displacement and strain distribution of F3C (Case C, 20℃, martensite) when loading displacement is equal to 1.5mm (d = 1.5mm) by using digital image correlation

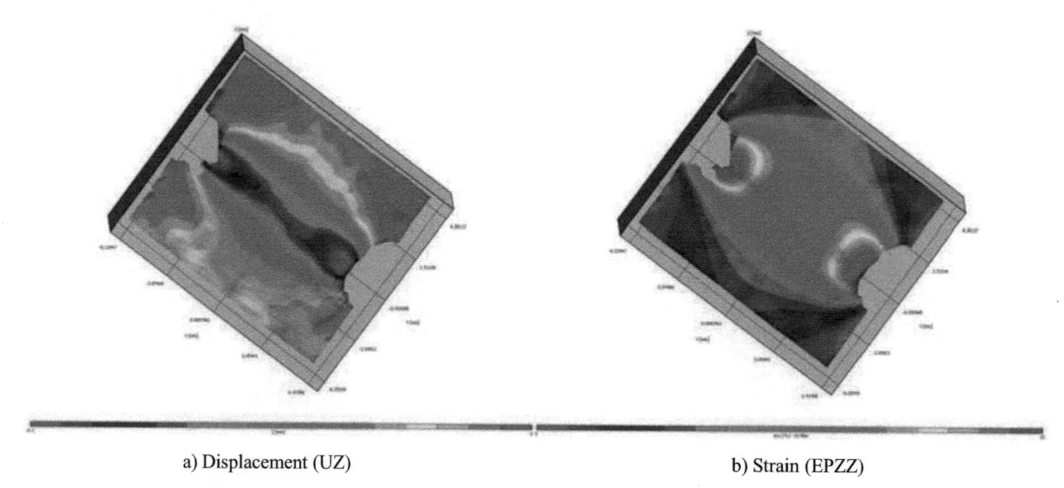

a) Displacement (UZ) b) Strain (EPZZ)

Figure B.32 Displacement and strain distribution of F4C (Case A, 200°C, mixed phase) when loading displacement is equal to 3 mm (d = 3mm) by using digital image correlation

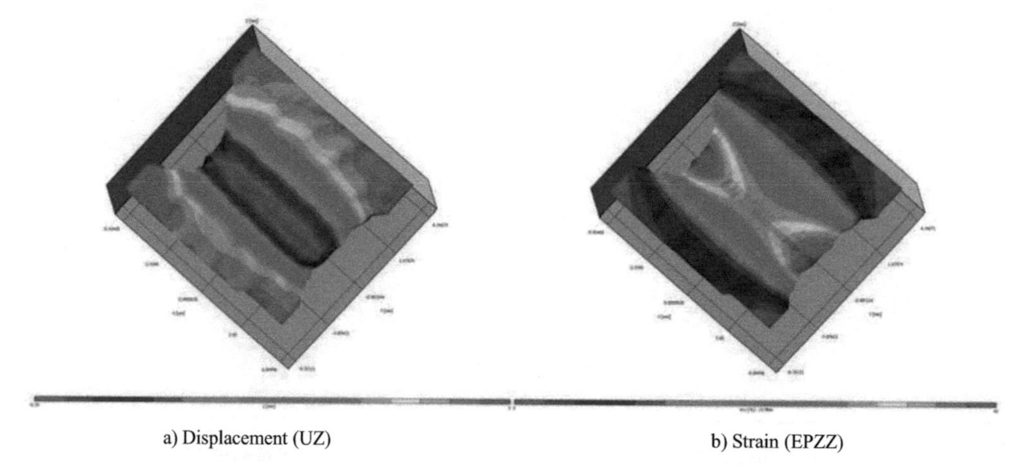

a) Displacement (UZ) b) Strain (EPZZ)

Figure B.33 Displacement and strain distribution of F5C (Case C, 300°C, austenite) when loading displacement is equal to 3 mm (d = 3mm) by using digital image correlation

Appendix C Example of Multiphase Program in CAST3M

```
*******************************************************
*                    --- MAIN ---                     *
*+++++++++++++++++++++++++++++++++++++++++++++++++++++*
* - Phase1: Martensite; Phase2: Austenite             *
* - Elastoplastic two-scale model                     *
* - Kinematic hardening                               *
* - Nonlinear material caracteristics                 *
*******************************************************


*++++++++++++++++++++++++++++++++++++++++++++++++++++++
*                     PREARATION
*++++++++++++++++++++++++++++++++++++++++++++++++++++++
*
*  The mesh
*  The temperature filed
*  The phase fraction field

OPTI DIME 2 ELEM QUA4 MODE AXIS;


*++++++++++++++++++++++++++++++++++++++++++++++++++++++
*                  READ INPUT DATAFILES
*++++++++++++++++++++++++++++++++++++++++++++++++++++++
```

```
OPTI REST 'FORMAT' 'Temperature.dat';
REST 'FORMAT';

OPTI REST 'FORMAT' 'Phase_fraction.dat';
REST 'FORMAT';


*++++++++++++++++++++++++++++++++++++++++++++++++++++++++
*                  MATERIAL CARACTERISTICS
*++++++++++++++++++++++++++++++++++++++++++++++++++++++++
* Evolution of thermo-mechanical propritities of 15-5PH

* Phase1
...
* Phase2
...


*++++++++++++++++++++++++++++++++++++++++++++++++++++++++
*      PARAMETERS OF TRANSFORMATION PLASTICITY MODEL
*++++++++++++++++++++++++++++++++++++++++++++++++++++++++
* Different thermal strain between two phases at 20℃
...
* Yield stress of Austenite
...


*++++++++++++++++++++++++++++++++++++++++++++++++++++++++
*               PARAMETERS OF DAMAGE MODEL
*++++++++++++++++++++++++++++++++++++++++++++++++++++++++
```

```
*Threshold damage strain (depend on temperature)

...
* Crack strain (depend on temperature)
...
* Critical damage (depend on temperature)
...
* Coefficient damage (depend on temperature)
...
*++++++++++++++++++++++++++++++++++++++++++++++++++
*           DEFINITION OF MATERIAL AND MODEL
*++++++++++++++++++++++++++++++++++++++++++++++++++
* Model of Phase1
MODE1 = MODL MAI1 'MECANIQUE' 'ELASTIQUE' 'PLASTIQUE'
'CINEMATIQUE' 'PHASE' 'PHA1';

* Model of Phase2
MODE2 = MODL MAI1 'MECANIQUE' 'ELASTIQUE' 'PLASTIQUE'
'CINEMATIQUE' 'PHASE' 'PHA2';

* Mixed model
MODEX = MODE MAI1 MELANGE PARALLELE (MODE1 ET MODE2);

* Material of Phase1
MATE1 = MATE MODE1 'YOUNG' EYOU1 'NU' COEF 'SIGY' ESIGY1 'ALPH'
EALF1 'H' HARD1 'RHO' MVOL;

* Material of Phase2
MATE2 = MATE MODE2 'YOUNG' EYOU2 'NU' COEF 'SIGY' ESIGY2 'ALPH'
EALF2 'H' HARD2 'RHO' MVOL;
```

```
* Mixed Material
MATEX = MATE MODEX 'PHA1' EPHA1 'PHA2' EPHA2;


*++++++++++++++++++++++++++++++++++++++++++++++++++++++
*                BOUNDARY CONDITIONS
*++++++++++++++++++++++++++++++++++++++++++++++++++++++

...

**----------Thermal loading-------------
P_TIME = P_TABLE . TEMPS;
P_TEMPER =- P_TABLE . TEMPERATURES;
CHAT = CHAR 'T' P_TEMPS P_TEMPER;


*++++++++++++++++++++++++++++++++++++++++++++++++++++++
*             CONVERGENCE CRITERION
*++++++++++++++++++++++++++++++++++++++++++++++++++++++

...

*++++++++++++++++++++++++++++++++++++++++++++++++++++++
*          CALL CLACULATION NONLINEAR
*++++++++++++++++++++++++++++++++++++++++++++++++++++++
OPTI DONN 'solveurnl.dgibi';
```

```
************************************************************
*                                                          *
*                  --- SUBROUTINE 1 ---                    *
*                                                          *
*                   'solveurnl.dgibi'                      *
************************************************************

*++++++++++++++++++++++++++++++++++++++++++++++++++++++++++
* DEFINITION OF INITIAL VARIABLES
*++++++++++++++++++++++++++++++++++++++++++++++++++++++++++

* Initial displacement: U(0) = 0
…
* Initial stress: SIGM(0) = 0
…
* Internal force: FINT(0) = 0
…
* Initial strain: EPS(0) = 0
…
* Initial plastic strain: EPSP = 0
…
* Initial transformation plasticity: EPTR = 0
…

*++++++++++++++++++++++++++++++++++++++++++++++++++++++++++
*             PROCEDURE OF NL CALCULATIONL
*++++++++++++++++++++++++++++++++++++++++++++++++++++++++++

* Loop J on the steps of time
REPETE TOTO JJ;
```

```
  J = J + 1;
  I = 0;

*---------------- Material properties on each step --------------

* Extract time and temperature (Step (J) and (J+1))
  TIM1 = EXTR TPS J;
  TIM2 = EXTR TPS (J + 1);
  TIMJ1 = MANU 'CHML' MODEX 'TEMP' TIM1;
  TIMJ2 = MANU 'CHML' MODEX 'TEMP' TIM2;

  TEMJ1 = CHAN 'CHAM' TEM1 MODEX;
  TEMJ2 = CHAN 'CHAM' TEM2 MODEX;

* Extract phase proportion J step
  PHA1A = PHASE1 . J;
  PHA1B = PHASE2 . J;

* Generate the alfa filed on J step
  ALF1A = VARI TEM1 EALF1 'SCALAIRE';
  ALF1B = VARI TEM1 EALF2 'SCALAIRE';

* Generate the alfa filed on J+1 step
  ALF2A = VARI TEM2 EALF1 'SCALAIRE';
  ALF2B = VARI TEM2 EALF2 'SCALAIRE';

* Generate Material properties filed on J+1 step
  CHAAT = MANU 'CHML' MODE1 'T' TEMJ2 'RIGIDITE';
  VMATA = 'VARI' 'NUAG' MODEI MATE1 CHAAT 'RIGIDITE';
```

```
CHAAT = MANU 'CHML' MODE2 'T' TEMJ 2 'RIGIDITE';
VMATB = 'VARI' 'NUAG' MODE2 MATE2 CHAAT 'RIGIDITE';

* Generate stiffness of material on J+1 step
VRIGA = RIGIDITE MODE1 VMATA;
VRIGB = RIGIDITE MODE2 VMATB;

* Addition of boundary conditions and the matrix of rigidity
KTA = VRIGA ET COD1;
KTB = VRIGB ET COD1;

* Coupling wint damage
…
* Material proporties
…
*++++++++++++++++++++++++++++++++++++++++++++++++++++++++
*                    RESIDUAL CALCULATIONL
*++++++++++++++++++++++++++++++++++++++++++++++++++++++++
* Calculate the FEXT
FEXT = TIRE CHAF TIM2;

* Give calculating infor
MESS 'time step No.'J' time = 'TIM2;

* Increment of the thermal strain: A, B, Mixed (type: CHPOIN)
DELT_THA = (ALF2A * TEM2) - (ALF1A * TEM1);
DELT_THB = (ALF2B * TEM2) - (ALF1B * TEM1);

* Calculate thermal strain
…
```

```
* Thermal strain of Austenite
...
* Calculate the thermal stress from thermal strain
...
* Homogenizing stresses
...
* Calculate increment of the thermal force from thermal stress
  FTHE = BSIG MODEX SIG_THM;

* The residual force
  RESID = FEXT - FINT + FTHE;


*+++++++++++++++++++++++++++++++++++++++++++++++++++++++
*             BANLANCE SOLUTION OF NODAL FORCE
*+++++++++++++++++++++++++++++++++++++++++++++++++++++++
* Loops to solve Fext = Fint
  REPETE BLOTO 1800;
    I = I + 1;

* Quit the loop in cas of non convergence
    SI (I > 1800);
      MESS 'non convergence on 1800 iterations';
      QUITTER TOTO;
    FINSI;

* Calculation of displacement increment
    DELTAU = RESOU KTD RESID;

* Calculation of cumulated displacement
```

```
    DPL_J2 = DPL_J1 + DELTAU;

* Calculation of the bonding force
    FLIAIS = REAC COD1 DPL_J2;

* Extraction of cumulated strain (total)
    EPT_J2 = EPSI (MODE1 et MODE2) DPL_J2;

* Move out the thermal strain (only mechanical strain)
    EPS_J2 = EPT_J2 - EPS_THA - EPS_THB;

* Remove TRIP
    EPS_J2 = EPS_J2 - EPS_TPA - EPS_TPB;

* Calculation of stress and flowing variables on i+1
    SI ((J EGA 1)) ;
        CHMM = COMP MODEX
          (TIMJ1 et TEMJ1 et EPS_J1 et VMATM et VMATA et VMATB)
          (TIMJ2 et TEMJ2 et EPS_J2 et VMATM et VMATA et VMATB);

    SINON;
        CHMM = COMP MODEX
          (restab. all.(J-1))
          (TIMJ2 et TEMJ2 et EPS_J2 et VMATM et VMATA et VMATB);

    FlNSI;

*---------- Calculate memorry effect ----------
...
* Extract and save the stress
```

```
    SIGG = EXCO CHMM (EXTR MODEX CONTRAINTES);
    SIGMM = REDU SIGG MODEX;

* Calculation of the internal force on i+1
    FINT = BSIG MODEX SIGMM;

* Calculation of residual force on i+1
    RESID = FEXT - FINT + FLIAIS;

* Calculation of the criterion ||Resid|| / ||Fint||

    NR = MAXI ((PSCA RESID RESID (MOTS FR FZ) (MOTS FR FZ))**0.5);
    NF = MAXI ((PSCA FEXT FEXT (MOTS FR FZ) (MOTS FR FZ))**0.5);
    CRIT1 = (NR / NF );

* Message of advanced calculation
    MM = MAXI EPSS1 'ABS';
    NN = MAXI RESID 'ABS';
    MESS ' Iteration | Criterion | Max strain | Max Resid';
    MESS '    'I '    'CRIT1' 'MM' 'NN;

* Stop of the procedure if convergence
  SI (CRIT1 < AR);

*++++++++++++++++++++++++++++++++++++++++++++++++++++++++
*                CALL TRIP CALCULATION
*++++++++++++++++++++++++++++++++++++++++++++++++++++++++
```

...

```
*+++++++++++++++++++++++++++++++++++++++++++++++++++++++
*                 CALL DAMAGE CALCULATION
*+++++++++++++++++++++++++++++++++++++++++++++++++++++++
...
*+++++++++++++++++++++++++++++++++++++++++++++++++++++++
*                CONVERENCE AND SAVE RESULTS
*+++++++++++++++++++++++++++++++++++++++++++++++++++++++
...
      MESS '                                        ';
      MESS '*** Convergence on 'I' iterations ***';
      MESS '                                        ';

      QUIT BLOTO;

    SINON;

* Replace increment U1 = U2 for next lOOp
     DPL_J1 = DPL_J2;

    FINSI;

  FIN BLOTO;

FIN TOTO;


*+++++++++++++++++++++++++++++++++++++++++++++++++++++++
*                   OUTPUT DATA
*+++++++++++++++++++++++++++++++++++++++++++++++++++++++
...
```

List of figures

List of tables